TEMPUS
Oral History
SERIES

voices of
KENT HOP PICKERS

Emma and Nellie Hatherton.

TEMPUS
Oral History
SERIES

voices of

KENT HOP PICKERS

Compiled by
Hilary Heffernan

TEMPUS

First published 1999
Copyright © Hilary Heffernan, 1999

Tempus Publishing Limited
The Mill, Brimscombe Port,
Stroud, Gloucestershire, GL5 2QG

ISBN 0 7524 1130 6

Typesetting and origination by
Tempus Publishing Limited
Printed in Great Britain by
Midway Clark Printing, Wiltshire

The set gets together on the last day of picking.

contents

The Napp family in 1955.

acknowledgements

My sincere thanks go to the many contributors of stories and photographs for this book. It has been a real pleasure meeting and talking with them all. I was invited to many homes, inundated with letters, even essays, and there has been a regular stream of visitors to my door generously bringing their memories in response to my request for hop-picking stories. My earnest thanks go to everyone who contributed towards the compilation of this book in any way and whose name I may have inadvertently omitted. I was overwhelmed by the sheer number of ex-Hoppers wanting to recount treasured reminiscences and their nostalgia for happy days gone forever. Many added that they have not told me all their stories and have many more where they came from. I'd love to hear them. All the pictures are from private, previously unpublished collections and I am grateful for their loan. I warmly dedicate this book to past and present Hop-Pickers of Kent.

Hilary Heffernan

introduction

These reminiscences are written as Hoppers told their stories. Some were recorded, others typed directly into the computer while being told, a few were written down as related. Some contributors rang or called a second or third time to add to their fund of stories.

Although the hop-picking season proper would not start until autumn, preparations and confident expectations began a good six months earlier. Patricia White says, 'The first requirement of a hop-picker is the Hopping Box.' All over London and surrounding areas barrows were being made out of strong wooden boxes or teachests, 'cannibalized' pram wheels were brought into service and fixed onto the box base, while two stout pieces of wood or broom handles served as barrow handles. The 'Opping Box was an important piece of equipment, carefully stored out of the way but easily accessible until Departure Day. Clothes no longer fit for everyday use were put in the box to be worn while hopping, together with dried and tinned foods, plus the essential oil lamp. In a world where unpaid rent meant eviction with your few worldly goods out beside you on the pavement and no 'welfare' to call on, every means of raising money was exploited. Returned bottles earned 1d each, paper rounds were pursued in all weathers, old boxes chopped to make bundles of fire-lighting faggots were sold for 2d a bundle and many other jobs were undertaken to earn cash to keep the wolf from the door. But paramount among all these, because this was the nearest they came to affording any sort of a holiday, was hop-and-fruit picking. The camaraderie and capacity for extracting every ounce of enjoyment out of the worst of incidents shows the marvellous spirit of the day and great courage under trying conditions.

Everyone recalled something special about those pre-hopping months. Far from damping their keenness, there was a satisfying inevitability about knowing they would soon enjoy a country holiday and add to their fund of stories. Their excited expectations sustained them through a year's hard graft and grind at home and work. Late August and early September saw an annual City exodus as hoppers travelled down to Kent in droves by lorry, bike, 'Shanks's pony', the Hoppers' Special steam train or any other available means, keen to enjoy anything from three to eight weeks' break in the countryside away from the dirt and noise of everyday living. It is believed the length of school summer holidays reflects the length of time parents kept their children off to go down to the hopfields but some still missed extra weeks of schooling to continue fruit or potato picking.

Labour was hard, the hours long, facilities were primitive and overcrowded, but this was a holiday and everyone was determined to enjoy themselves. Some crossed over from Essex, others already lived in hop-picking country. As a child Mrs Gwen Smith lived in East Peckham and picked in Whitbread's hop garden nearby. Lily Dicken lived on a hop farm and really earned her lace-up boots and black stockings by the end of the season. Iris Deick would have liked to go with all her friends, but her parents wouldn't consider hop-picking as they thought it was too rough.

A few found the basic facilities, crowded living conditions and insects more than they could take: straw mattresses were prickly to sleep on; fingers tasted bitter from hop juice accumulated while picking; there were mice, nettles, bruises and burns, scratches and cuts; but for most these all went for nought compared with the long-lasting friendships formed, cheerful company and pub evenings or singing around campfires under the stars.

Sometimes there was romance and matrimony. Always there were jokes, teasing, the ability to laugh at oneself and the integral honesty of poor people who stuck up for each other and in adversity rejoiced in the good fortune of others. Jealousy and rancour were quickly forgiven and forgotten. Despite poverty there was a real pride in managing with what they had, and who can blame them if sometimes they scrumped a few apples or the odd duck? Not for them, living off State handouts; independence was too precious to be beholden to others. Despite all this there was no vandalism, no real crime. At home, doors were left on the latch with keys dangling on bits of string through letter boxes, potentially accessible to any passer-by. No one stole from neighbours' precious few belongings. The poor respected the poor. There was a strong sense of security because everyone knew their neighbours and relaxed with each other. Old people sat out on the pavement in favourite chairs; passers-by greeted them by name, stopping for a chat. There was no need to feel lonely; unlike today when people living next to each other for years are strangers when they meet. There was discipline, too. A neighbour seeing a child misbehaving quickly responded, 'Stop that or I'll tell your mother!' – a strong enough threat in those days to be fully effective. Not that children were angels: despite being forbidden certain activities or dangerous areas, the adventurous spirit soon took over whenever opportunity beckoned and many a scrape is recorded here to show that children were every bit as full of spirits as today's offspring.

The time of hand-picked hopping is gone but not the warm wonderment of it all. Memories of happier days are clear and detailed. Many contributors, reading my previous book *The Annual Hop: London to Kent*, were prompted to add their own stories to this fund of hopping tales. Most wrote about the daily routine of the hopfields and some descriptions are included as a reminder of those early mornings when the first, dew-laden bines soaked the pickers as they reached for the juicy hops and raced to finish their aisle before the family next to them. We remember the tallymen whose job was to be fair but strict in their measuring, yet wary of pickers plumping up the hops in their basket to look more, or who hid generous supplies of leaves and bits of bine at the bottom of their basket (known as 'dirty pickers'.) Some farmers provided accommodation of comparative luxury to which the same families returned, year after year. Others offered tents, leaky corrugated iron huts or even animals' winter sheltering, which needed cleaning out before the family could settle for their stay. 'Facilities' varied from finding

The Wood family.

comparative privacy behind a bush or using a plank precariously balanced across a lime-filled hole, to the provision of proper chemical toilets. Personal hygiene was usually achieved standing behind a blanket suspended across the hop hut while using an enamel washbowl of water and Lifebuoy or Coal Tar soap. Even seasoned campaign soldiers found conditions on some hop farms primitive beyond belief.

They called them hop gardens or hopfields and hoppers tend to stick religiously to the name by which their farm was designated, so you will find both titles here. At least one lady was still picking at ninety and when their time came, some asked for their ashes to be scattered in 'their' hopfield, faithful till the last. Some were born at the hop farm and picked every year until the advent of mechanization in the mid-

fifties. Hops were sometimes included in bridal bouquets and the Hop Queen wore a hop crown for her day of glory at the end of a season. Whether they picked for one year or fifty, hoppers remain constant to their memories. I have tried to bring out the humour, happy satisfaction, warmth and the pioneering spirit of those hop-picking times. Nostalgia for what the majority remember as their happiest days and best holidays ever has not diminished with the years.

CHAPTER 1
Preparation and Anticipation

Arthur Smith's family outside their hopping hut.

Singing Along

We lived at East Peckham near Whitbreads. I'm eighty-two now but when I was a girl I went picking with my mother and sister. I'd look out of my bedroom window and watch the lorries come down from London, the people with all their pots and pans. They made their huts very nice, with curtains and everything you could think of. We didn't relish the early morning starts in the mist, pulling wet bines. We sat on a box, picking into an upturned umbrella and listened for the lollipop man shouting 'Large ha'porths and big penn'orths'. If we picked a lot of hops we were allowed some sweets. We'd go by the hoppers' huts and watch them making a big fire on bricks and faggots for their pots of stew.

Gwen Smith

Invitations and Tickets

Just after Christmas our mum used to write off for the hop-picking. We had a big wooden box on four pram wheels and two wooden handles with a wooden lid. We put in all our warm clothes, wellington boots, pots and pans, kettle…everything we needed for camping out for three weeks. We went to Highwoods for twenty years. The farmer sent our rail tickets and the letter, and it stood behind the clock from early summer. The excitement when that came was unbelievable. We had six of us jumping up and down, keep getting the letter down, reading it. We caught the New Cross to London Bridge train for the Hoppers' Special.

Our two brothers went in the Guard's van with the box. One carriage mum wouldn't let us go in: the old man had a bucket in there, then threw the piddle out of the window. Our mum started hopping from Deptford years before and most of the family were christened at St Paul's in Deptford High Street. When the rest of us came along they moved to Church Street, Charlton.

Jean Pilbeam (née Bird)

Windy

We all trudged up to London Bridge station pushing the hopping box to catch the Hoppers' Special. We lived in Dockland. One year we went down by open lorry with chairs and benches in the back for seats. The trouble was the driver, most of the men and some of the women were in the pub until lunchtime closing so we were very late setting off. That year we went to Hawkhurst which was a long journey, so it was nearly dark by the time we arrived. We'd never been there before. Our hut was called 'Gone With the Wind' – aptly named. It was full of holes and cracks in the door and we had to settle in by candlelight. Pre-war we went to Yalding and sometimes Whitbread's farms but the last years we went to East Peckham.

Joyce Rogers (née Tibbott)

Pushing the Charabanc

Grandfather Castleton lived in Graham Road in Welling in one of the old cottages where the school is now and Aunt Flo and Aunt Edie Love lived in Welling too. We picked at Manor Farm in Laddingford which is about three miles from Paddock Wood. At first it was owned by Mr Banks, then by Mr Chambers. I started going in 1924 when I was only a few weeks old. I was born on 24 February. We used to go down with Mr Joe Breeds in his charabanc with its solid tyres. They've improved the roads a lot since then but sometimes, on the way home, we all had to get out and push it up the steep hill at Wrotham. On the hopping trains if they didn't want to pay for the kiddies they put them under the seats and made them hide behind the ladies' long skirts.

Arthur Smith

'The Letter's Come!'

It took all day to go from Deptford to Horsemonden to Mainwaring's Farm. I went there several years as a kid and loved it. I was one of the fortunate kids who used to go for seaside holidays but I much preferred the hopping holiday! I had all my mates and I could run free in my old clothes. I loved it. You knew when hopping time was approaching because the local women all started asking each other if they'd received their letter from the farm inviting them to go down in September. If they hadn't received their letter and everyone else had, they'd start worrying in case they weren't wanted. The local Woolworths stocked up with all the cooking utensils such as the 'opping pot. After the pickers came home the local economy was raised a few notches and there were

plenty of hopping apples about. When the hopping season was on it was like a desert around Deptford until the families came back from Kent or Sussex. I know our teacher at my school used to go potty; we'd come back after the summer holiday, then within a few weeks the kids would be away again just as he was knocking them into shape, literally.

Mr T.W. Ovenell

Long Walk

The letter from the farmer with a family travel ticket would arrive and we'd pack our wooden box on wheels and label it 'Beech Farm, Marden' ready to push the five miles from Camberwell to London Bridge for the old steam train, the Hoppers' Special. At 4 a.m. it was no fun and there used to be queues when we got there. The train was full of hoppers. The farmer collected our luggage when we arrived, but we had to walk another five miles to the farm.

Brian Smith

Yearly Exodus

Our preparations began in June when the train tickets from the farmer dropped on the mat. The first thing we got ready was a large packing case with pram wheels and handles. We put in everything we were taking with us – pots, pans, bedclothes, our own clothes such as they were, and anything we might need for our five weeks' yearly exodus into the country. We got up at three o'clock in the morning and walked from Vauxhall, where we lived, to London Bridge station, to catch the Hoppers' Special at 6 a.m. I was about five years old then and that was an adventure itself. We were met at Paddock Wood by huge farm carts to take us to Woodfalls Farm in Yalding.

John E. Meinke

Whitbreads

I loved every minute of our ten years' hopping at Whitbreads. I still remember my mother applying and her four sisters each year, and waiting for the postman every day for the letter to say which farm we'd be picking on. When it arrived we were so happy. It was as good as winning the Pools in those days. My mum would say, 'Here's 2s. Go up to the oil shop and get me a set of hopping pots.' They started off all lovely and silver shining but by the end of three weeks cooking on open fires they were really black. But our mum cooked delicious meals in them. We lived in the East End so you can imagine three weeks in Kent were a real treat.

Mrs R. Vaughan

Cranbrook

Dad would go down to Chrisp Street Market to borrow a big wheelbarrow. He'd bring it home and load it up with all the stuff mum had ready for hopping, then put we three children on top of the luggage. With Dad pushing, they'd walk all the way from Poplar to London Bridge with about 150 other families. At the station all the children ran ahead through the gate as we couldn't afford their fares. Home for the next month or more was to be a ten-by-ten-foot brick hut at Cranbrook. It was our dining room, bedroom and kitchen for as long as it took to strip the fields of hops. A picker may earn £20 but that was like winning the lottery then and Mum made sure we'd all have a good Christmas.

Mr C. Norman

Our Only Holiday

We looked forward to the hopping as it was the only holiday we got and we liked to see how many of the people we already knew were coming. We collected all the tin plates, cups, primus, pots and pans, old enamel washing-up bowls and all the other hopping things we'd stowed away, plus lots of tinned stuff. I was about thirteen so was too old to take toys. We had an old tin trunk and the usual brown strong suitcases and boxes to sit on. Our neighbours took us for two years, then we went by train. We took any old clothes; not many kids wore trousers then, not me. We had old overalls, skirts, macs and wellies. Dad stayed at home so he could go to work, as my sister did later on. When I started work at fifteen I did my week's holiday there.

Enid Styles

Rosina's mum, Rose Elliott (left), and her cousins.

Nine Children

We lived in Stepney Green and my mother and father were Lilian and George French. I was one of nine children. My dad's sister, Daisy, came hopping with us, sometimes at Yalding but we went to several farms at Cranbrook, Paddock Wood, Wateringbury and Goudhurst. We lived in Dagenham then. My first year picking with mum and dad was in 1947 and from what I remember, we went to Goudhurst by train from Bromley. At one time we went by road to Mr Barden's farm.

Doris Turner (née French)

Long Wait

We waited all summer for the letter from the farmer to say if we could go. We were always glad when we knew it was okay. I was six when I first went with my Aunt Rose and my two older brothers. It was September 1939 and war had just been declared.

Graham C. Turner

The Letter

Gran used to arrange for the visits to the hopfields. She'd eagerly wait for the letter arriving each year so she'd know the family could all go. Our neighbours would ask each other if they'd got their letter yet. My maternal grandmother, Ethel Hatherton, went hopping most years until she died at eighty-five.

Joyce Rogers

Regulars

My family were regular hop pickers. I remember helping Mum to pack all our things in the tea-chest from about the end of July. We travelled to Otford by Jim Grey's open lorry covered in a green tarpaulin, though one year we went by horse and cart. My mum, Lilian, is ninety-four.

Stan Dalton

Mates

I lived in Neate Street, Camberwell. My mate Charlie Ray came from Canal Place. His house backed onto ours. This was at the beginning of the war. We were about seventeen then. His family always hop picked at Horsemonden. The hut was only corrugated iron but when his mum had finished cleaning it inside, it was very nice. We got faggots from the end of the field, a long way from the huts. Straw was already there in bundles when we got there. The faggots were put down on the dirt floor then straw on top, which made a very good bed for three or four or us. We'd a small table and paraffin lamp and that was all. The firelight meant we didn't use the lamp much.

Mr C.H. Wiggins

Feather Bed

I spent many years down hopping with my grandmother as a young child. That was our holiday. We went by lorry to Mr Hubble's farm at Brenchley, three miles from Yalding. We took everything we needed, even tables and chairs, and grandmother even took her feather mattress. We had to make the beds up on a platform at the back of the hut. There was straw ready in the hut for us to use for filling the extra mattress covers and pillow cases. We put lace curtains up at the door and the big box granddad had made to carry all the china and clothes in was converted into a table with a drop leaf so the food was safely stored inside.

Mrs N. Burnby

Hopping Family

We went picking down at Horsemonden from 1915, all the family. In fact I was nearly born down there. I'm seventy-nine now and I still work at the National Theatre in London.

Harry Henderson

Small's Farm

Most of our bits and pieces went into an ottoman box, things like a mattress tick which we'd fill with straw and blankets, of course, because the huts were cold and draughty. Many things were left down there year after

Shirley Whiting with her mother, Sylvia Hines and Doreen.

year: a heater, a small table and some chairs, pots. We used to go down from Anerley station to Small's Farm, Cowden, near Penshurst.

Q. Moody

Old Memories

My family picked at Freraliss' Mill Farm in Wateringbury for many years. I was six weeks old when I first went in 1925 and we went every year until 1939. Mill Farm hasn't changed much. I often visit it to revive old memories and the butcher's is still the same. Aunt Martha, Mum's friend, ran a café under the boathouse.

L.C. Presley

Mayor of Stepney

We lived in Limehouse. My grandfather was Mayor of Stepney and died in 1934. His funeral was one of the biggest in the East End. The cortège went along Salmon Lane, all lined with people. Even Clement Attlee, Ernest Bevin and Robert Mellish came. When I was six my Aunt Mag and uncle took me hopping. That was 1947. Uncle borrowed a lorry and we packed everything we needed into it and drove down before the hopping started so we could prepare our three rooms as we had such a big family. Everything we'd need for the next two weeks was packed into tall cane baskets. That included crockery, pots and pans, clothes and, most important, a mattress cover for the straw for our bed.

Mrs M.E. White

Regulars

I came from Greenwich and my aunt was a regular hop picker. Hopping was a summer holiday for most of us. My uncle used to borrow a covered van and we loaded up with everything – clothes, bed linen, pots and pans, tables, stools and even a small dresser and the all-important oil lamp. I can see it now, all loaded up like a removal van with my aunt, cousins, me and my mother and the family dog all in the back ready to go. My uncle took two hours to drive down. We used to sing songs and cheer other hoppers as we passed them.

Clive Gwyer

Smuggler's Farm

My husband was brought up in Covent Garden and for a many years he went hop-picking with all his family at Smuggler's Farm, Goudhurst.

Mrs N. Daley

Families Shared

We took everything but the kitchen sink. Families sometimes shared hiring a van or a lorry. We went mainly to Paddock Wood, Lamberhurst and once to Whitbreads.

Irene Parton

Thick Mud

September, we went hop-picking. I'd get everyone down on my lorry but would have to be back at work next Monday so I'd only be there over the weekend, which was good again as the pubs down there were always packed but you had a drink and a good time. You had to pay 2s 6d on the glass for the first drink but you got it back when the pub closed. Rain was the worst weather you could have at hopping. The mud was thick. I felt very sorry for the families with small kiddies in prams, pushing and pulling, up to your boots in mud.

This was your two weeks' holiday in those days and all those hoppers loved it.

C.H. Wiggins

Mending Shoes

We came from Beaconsfield Road in Mottingham and I went to Mottingham County Secondary School in Ravensworth Road. They demolished it in the 1990s and it's all new houses now. Ours was a council house. Mum never went to work and Dad was a conductor on the buses. Shirley Smedley was our next door neighbour and did we get up to fun and games! She emigrated to Canada about 1961. Mum was a dressmaker and made all my clothes and my sister's and did work for customers. She formed a 'Make and Mend Club' in Mottingham. Dad mended all our shoes. I still have his last. He used slabs of hide and put Blakey's studs on the toes and heels to make them last longer. We travelled with our neighbours next door, or on our own. It took about an hour to get down to Chart Farm, Seal Chart, near Sevenoaks. Neighbours went from all over our Mottingham Estate.

Enid Styles

Like Moving House

Our mum started packing tins of food and household things weeks before we actually set off. It was almost like moving house. I think the best part for us kids was seeing the sweet jar gradually fill up! The furniture lorry we hired always turned up very early to take us to our farm. We all sat in the back with our sandwiches and flasks. A lot of our bigger items were left down there year after year. We lived at Woolwich and only had to go to Faversham but didn't get there until the afternoon. The huts were wooden with corrugated iron roofs.

Mrs C. Mortimore

One Happy Family

Hopping was cancelled for a few years when Dolly and Joan, my twin sisters, were born in 1940. But when we started again there were three families in our block of flats went and we all waited eagerly to see who got their letter first. There was our family, The Tylers (that was Fred, Vi (Aunt Bob), Freddy and Rose, Georgie and granddaughter Janet) and the Lynches: Jim, Rose, Joan, John, Terry and Carole. We were sent assisted train tickets and our hut and bin numbers with the letter. We left Vauxhall Street at about 5 a.m., pushing hopping boxes and carts through Lambeth Walk (which was deserted) to London Bridge station. If we were lucky we'd find an empty compartment and Aunt Bob got us kids to hang out of the window and do our whooping cough act to keep it to ourselves.

Bob Orris

Six Weeks in a Tin Hut

My family, that's my parents, myself, my brothers and my sister, lived in McNeal Road, Camberwell, SE5. Apart from my dad we all went hopping every year at a farm in Yalding. Not only us went but the whole street, too: almost all our neighbours. You can imagine a whole street going picking hops! But while we were away all that time I can't remember anyone ever getting burgled or robbed. We took everything we could carry and went by lorry to the farm for the whole six weeks' school summer holidays. We lived in a tin hut. My dad was working at the time but he never missed coming down to the farm every Friday night to spend the weekend with us.

W.D. Leman

Welcome Holiday

I'm ninety-three years old and have twenty-seven grandchildren, half of each. We lived in The Cut in Lambeth and went to Yalding at East Peckham for years. My mother took us young ones for our holidays. We caught the train at London Bridge. The farmer waited for us at Yalding station. He'd take our carts and we'd all walk the rest of the way singing. It was a real treat.

Mrs M. Sanders (née Ede)

Friends and Neighbours

Our friends the Sullivans and our family always went hopping at

Wateringbury together. We all lived at Dockhead. Instead of evacuating when war started, my nan took her six children hop-picking. There were eight Sullivans, the father Din (Dennis), Alice, the mother, and their children Eric, Dinny, Con, Maureen, Terry and Pat. At first we were the only families, but a few years later the Wrigleys, Hornes and Philpots came too.

Irene Hans and Debbie Wetheridge

Life was Good

There was no privacy in the hut and we didn't have any baths. We all just lived together so had to do the best you could. Mum did the shopping in Sevenoaks on Saturday afternoons. We had marvellous views over Sevenoaks and the Weald of Kent. When the weather was good life was good. We met all different people, especially home dwellers or day pickers. I go camping now so still enjoy the open air life.

Enid Styles

Fare's Fair

Coming from Deptford I was right in among all the hop pickers. We mainly went down by train, the Hop-Pickers' Special. If some of the mothers were a bit hard up for the fare they put their youngsters under the seat so they didn't have to pay. Later we went by lorry and it took us all day to go from Deptford to Horsemonden.

T.W. Ovenell

Mr and Mrs Orris, Jimmy, the twins and a friend at Goudhurst in 1953.

Hoppers' Special

Our family lived just off Tower Bridge Road. Joan, my older sister, lived in Queens Buildings and I babysat for her every Friday night after school. I got 2s. Every time we came back from hopping I had to attend Spa Road cleansing station because of the nits we picked up there. We went to Ploggs Hall Farm at Paddock Wood with my Mum and Dad from about 1949 to 1956. We travelled by train but our bits and pieces went in the van with our relatives.

Patricia Facer (née Wickham)

The Wetheridge family.

Homestall Farm

Mother took us picking at Mr Bones' Homestall Farm. In 1952 she was asked by the formidable Mrs Porter at Swanscombe if she'd like to take the place of someone unable to go that year. She accepted, although dad had reservations about the whole thing. He came from Jarrow and didn't know about hops. My mother had picked with her grandmother in the early 1920s. We went off in a smelly, tarpaulin-covered coal lorry to Faversham and were given a corrugated tin hut, one room, no heating, one big bed and a straw pallet. We slept end to end. Mum cooked on a small primus or on the outside fire. It rained all the first week and when Dad came down, Saturday afternoon, we must have looked bedraggled; he insisted he'd take us all home. He was surprised we were enjoying ourselves despite the weather. We had freedom, fresh air and no danger. It was marvellous. As long as Mrs Porter didn't see us, we scrumped for fruit – fresh food was in short supply. We only saw oranges at Christmas. Some larger families were housed in large barns, complete with owls in the rafters. Dad eventually came round to the idea of an annual hopping holiday and took his two weeks' holiday for then. He wasn't the world's best picker but could rustle up a decent cuppa and crack a good joke.

Doreen Dillon (née Wallace)

Down to Goudhurst

We were born and raised in Bermondsey, all six of us, and every year we, our parents, any aunts and uncles who wanted and the dog went hop-picking at Three Chimneys Farm, Goudhurst. The first few years we went down by Hoppers' Special from London Bridge to Goudhurst but in later years we travelled in so-called luxury by lorry. The Hoppers' Special left early while it was still dark. There was a great crush of people all round the station with their hopping carts, bundles and bags. Half the children didn't have tickets. Even the cheap fares were too expensive for most. Most of the hopping boxes were just a box with four or two wheels on them.

Joan Jeffery

Paradise Street

I was born in 1933. We went hopping with several other families: the Cronins, Cheritons, Kasners, Woods and Hedgecocks. We all gathered outside 58 and 60 Paradise Street in Lambeth. The passageways were filled with tea chests, boxes and prams waiting for us to put them on the lorry for going down to the hopfield, but first the driver took all our families to the station to catch the Hoppers' Special at London Bridge or Waterloo. The trains were always so full children stood or sat on someone's lap. It was a long walk to the hop hut field from Marden station; turn right then first right again onto Long Lane, Indian file, all families in groups. To the T-junction, left again.

Then the oast houses, pond and conker tree and thirty yards to our hut field.

Jim Wood

Crush

There was a huge crush of people waiting to catch the hoppers' train from Bermondsey to Marden with my Aunt Mary and Uncle Jim. Everybody had hopping carts and luggage. When the gates opened we'd all make a sudden surge forward. We'd be pushing past the ticket collectors if folks had tickets or not. I was about eight years old in the 1920s.

Bob Richards

Last Laugh

We were a hopping family. I lived in Deptford and this was the only way to get a good holiday for the kids and a bit of money for Christmas. We picked at Marden and I think I was conceived there. Anyway I was born at the end of May in 1925. The farmer sent us a cheap railway pass to go from New Cross station but it was very early in the morning and we couldn't take a lot with us so we always got Uncle Bodger Darling to get a van. We got all the tables, chairs, beds, prams and kids in it. It always broke down. It was overloaded and we were always a long way from help. One day after all us kids had done a wee-wee in a little potty, Mum leaned out the back of the lorry and it just missed some men cutting the hedge. We all jeered and laughed, then

Jim Wood's family.

just up the road the van broke down and we all had to sit there while the men walked past, calling us all the dirty names they could think of.

Joan James

Tradition

Tradition was that each family kept their own hut each year unless a family gave up going, which was most unlikely as this was their holiday with a chance to earn some money. Huts were in pairs or fours with two-foot gaps between. Each group had a place for a cooking fire, fifteen or twenty feet away. If it rained the cooks always got wet. No such thing as lunch time, just a snack of sandwiches with beer or tea. In general it was the friendliness that made the hard work easier to bear. A van arrived with groceries at the weekends, selling groceries, fruit, occasionally chocolate or ice cream which had 'fallen off the back of a lorry'.

Jim Wood

Sacked

Some of the mums put their children in hopping sacks under their seat on the train so they didn't have to pay their fares. When the ticket man came round you'd hear them whispering, 'Shut up now or I'll belt you one.'

Nell Hearson (deceased)

Helping Hands

Hop-picking was a way to earn money pre-war. I was ten years old the first time. A neighbour of ours – her

Rosina Amis having her hops tallied.

husband had a terrible accident in the docks. She asked my mother if we could go with her. That was me and my sister; she was twelve. We had to leave at twelve o'clock at night to get a train from London Bridge. We pushed a big pram and a cart all the way. It was really exciting because I'd never been hopping before.

Mrs A. Hamlin

Overloads

Ours was a large family. We lived in Bermondsey which was a deprived area in those days. September was the month we looked forward to, especially the first week when we got the letter telling us the hops were ready to be picked. My mother was a great organizer and she'd hire a lorry for five or six local families who were going to the same farm in Marden. You can imagine what

it was like with six families, five in each, all bundled into a lorry on its way to the wilds of Kent.

Mr T.E. Fielder

Scene Change

My mother had always gone with her family to Kent where the hopping season was longer. The hop-tallying method was different to Hampshire. For the children, it was a release from narrow streets and concrete school playgrounds. After the war mother and her sisters were invited to join friends at Bentley, Hampshire and found the conditions far better than their Kent farm.

Rita Game

Southwark

I lived in Shalford House, Law Street, just off Tower Bridge Road near Hartley's jam factory. My mother and eldest sister both worked there at one time. My brother, Johnny, was at Barrow and Hepburns when he first left school. My oldest sister lived right at the top of Queen's Buildings. The toilets were shared between four families and there was no lighting on the stairs: I used to be terrified going up and down after dark. For years when I was young I thought going on holiday was going hopping. I didn't realize that going to Margate or Southwark was a 'proper' holiday.

The trouble was that when we returned to school for the autumn term we were a week or so after everyone else so all your class knew where you'd been. I went to Grange Road Junior Mixed just off Tower Bridge Road. We went hopping at Ploggs Hall Farm in Paddock Wood when I was a child between 1949 and 1956 with my Mum and Dad, my sisters Joan and Ann and my brother. We went by train but our bits and pieces were taken by relatives in a van. Our hopping huts had no electric but I can remember people saying Whitbread's had electric in theirs, which seemed very posh in those days.

Patricia Facer

Rosina Amis's mother cooking dinner.

CHAPTER 2
Country Living

Mrs Fagan and her sister-in-law, Doreen Brown, in the hopfields.

Mrs Fagan's sister.

red hot. They baked our Sunday roasts for us and all we had to do was put our hut number on it with a dolly pin and what time we'd like to collect it, so on Sundays we only had to cook the veg and the spotted pudding for afters.

Mrs W. Fagan

Sunday Roasts

We went with my mum to Mr Finn's farm at Chartham near Canterbury on about 3 September but we didn't start picking until about the 10th. We were there about six weeks. Our huts had two large bunks and we had to sleep head to toe. We put up curtains to make part of the bedroom. There was a stove to boil the water for the morning cup of tea but our food was cooked on open fires. The farmer always sent us plenty of faggots. We'd start picking about 6.30 a.m. when a tractor was sent to give us a lift to the gardens, or we could walk if we liked. It took about twenty minutes. The villagers baked our bread every day, which was

Palliasse Beds

Our family kept the same hop hut year after year. We used to distemper the walls so it looked more homely. Mr Banks let us leave our bits and sticks of furniture in the hut until the next year. When we arrived we'd collect straw from the farmer and have to make up our own palliasse for a mattress and collect faggots for the fire. We all got up at six, even the children, and whatever the weather we had to go down to the fields. Old Alf Hester used to wake us up by walking along the huts and rattling a stick all down the corrugated iron walls. You couldn't sleep through that! He came from Ruby Street in Old Kent Road. We'd have a bit of bread and jam before we left and on the way down call in at Brenchley's Provision Stores and get a loaf and some Dutch Edam. This was our day's food.

Arthur Smith

Frosty Days

Mostly the hoppers came down by train from London. The farmer met them at the station with a horse and wagon and took them and all their luggage to the farm. He put them into

Arthur Smith's family outside the top hut.

tin huts to live in. When the picking was over it was a grand sight watching them going back to London. While they waited for the train to come they'd be singing and dancing on the platform. When there were early winters or if we had frosts we couldn't go picking so we'd run off and play.

Mr F. G. Pearson

Marden

We picked at Marden. A horse and cart collected us from the station and took us to our huts at the farm. The farmer left bundles of straw by the huts for us to use for filling the mattresses and pillows. We had the same huts every year. While our aunt got the hut ready we kids used the time to check the orchards. I remember September 1939: the talk was of the war which had only been declared ten days before. Mornings were the worst time. Up at six, we washed in cold water, although sometimes we were lucky enough to have warm water. We had toast and cocoa for breakfast and had to be at our bins by seven. It was hard work for youngsters but, looking back, it was worth it. We really enjoyed ourselves.

Graham Turner

Lily Dickens takes a break.

Hot Toast

After toast and tea in the mornings made on the cookhouse fire it was up the field to pick hops. They were full of dew when the bine was pulled, but the sun seemed to come out later every day. It was wonderful. Mr. Watts made tea on a barrow when we stopped picking to have lunch. It was too far to walk back to the huts on the common. We'd eat sandwiches holding paper round them because our fingers were black from hop juice and they tasted very bitter. Our hops were measured twice a day into bushel baskets and entered in books by the woman clerk so we'd know how much we'd earned. Once a week the farmer opened his

office so people could have a sub if needed.

Mrs N. Burney

Hut Fumigation

As soon as we arrived mum set off a sort of stink bomb in the hut to fumigate it, then we'd get a fire going outside and put some kettles on for tea. A cart came round with bundles of faggots for our fires. Dad was still at home because of his job, but came down at the weekends. Mum sorted out the hut and we kids would run off to explore and fetch water from the pump.

C. Mortimore

Transported by Bike

I'm nearly eighty-four now but I can remember my mother taking me on the back of her cycle to go hop-picking at seven o'clock in the morning. In hard winters and frosts we couldn't pick and all the children would either sit on the bin or play horses which was when we tied a piece of string on each arm to make reins to hold and ran round the hop garden. When picking started the children were made to pick too. If they ran off their mothers ran after them and, when they were caught, would thrash them with a stick.

F.G. Pearson

Ethel Chandler's hut at West Malling in 1955.

Delicious Dish, Cold Fish

I've got such vivid memories of going to Mr Neame's gardens. My first time was about sixty-six years ago by open-topped lorry. There was Mum with four children, her parents and old friends of theirs called Christopher. We packed everything we'd need into tall cane baskets, including the important mattress cover which Mum stuffed with clean straw for the bed. Dad came down at the weekend and we all slept in the one bed. Dad's parents came down and they and my nine-year-old brother slept outside. Mum picked hops into the baskets we'd brought our things down in. Every so often farmhands came round to do a tally. Mum's hops were tipped into a bigger basket and he'd write on her card to show how many bushels she'd picked. Sometimes we children helped, but mostly we were off playing. The fish man came into the gardens selling cold fish and chips. Maybe that's why I still like cold fish. Another man came with cakes to sell, very likely stale. There was ice cream and, of all things, tomatoes. With all the fruit we ate the two deep trench latrines must have been in constant use. I've no idea how many people used them but I remember queuing outside for my turn. The Clinch's village shop we used was lit by oil lamps. I still remember the smell of paraffin, groceries and soap.

Mrs M. Murray

Inside Irene Crimins' hop hut in 1976.

Only Natural

For years my family went to Moat Farm at Five Oak Green owned by the Tolhurst family. My mother didn't like it much except for visiting so my brothers and I went with our nan and grandfather who was a pole puller. We kids had to pick so many umbrellas-full of hops before we could play. When my brother was about six weeks old we had so many visitors came down all the women slept sideways in our big brass bedstead and a curtain was drawn at the bottom. The men had to sleep in tents and the cookhouse. One uncle slept in a single bed at the end of the double bed and my dad fixed up a hammock for a cot for the baby and hung it over the single bed. Of course the baby wasn't wearing waterproof nappies as they'd never been heard of in those days (1934) so my uncle got a soaking and the air was blue. They passed the word all along the row of huts which only had thin partitions between, and gaps at the top. The families were all in hysterics – uncle's language was very colourful.

Irene Crimins (née Rene Thurbin)

Gold Lace Curtains

Our accommodation was quite good, even luxurious compared to some. We had brick-built huts with corrugated iron roofs and sliding doors. The bed was built into one corner and it was big enough for four people if they slept in

30

the usual manner, or up to seven or eight of us if we lay across the other way. We used straw mattresses but my mother always took a feather pillow for herself. She took gold lace curtains to hang round the bed and bits of carpet for the floor. Some people even hung wallpaper, but we were never that ambitious. The huts were built around an open-ended square with a central brick cookhouse of four fireplaces built round a central chimney. They had bars across the tops to hang our pots on. Everything we ate had to be cooked there so nothing was baked except the odd potato put in the ashes. The privies were across a small common. There were six of them, brick, built over a trench. There weren't any buckets and no individual cubicles, but there were different blocks for men and women.

Joan Jeffery

Mrs Fagan's sister does the washing.

What a Lovely Life

We had oil lamps, straw-filled mattresses, crockery, cutlery, slip mats and what we needed. The huts were in double rows so there were huts backing onto each other. It was quite comical listening to families chatting and calling out to each other. The mornings were cold and crisp in September and you could hear people talking outside which woke you up about half past six. We had a primus stove for boiling the kettle so we'd got water to wash, then maybe have a mug of tea and a slice of bread and jam. Mum made sandwiches or filled bread rolls for our lunches we'd go back to the huts for lunch if they were close by the field we were working in. After we'd pulled the last bine about five o'clock, we'd have stained fingers and muddy wellingtons so had to clean up. Mum lit a fire outside and started cooking supper – it might be sausage and mash or a nice vegetable stew. What a lovely life!

Doris Turner

Family Bed

Our "Opping Box' was wood and measured about 4ft long by 2½ft deep. It had a set of wheels and a handle taken from a cannibalized pram. On the morning the hop-picking season started all the family and the hopping

Leslie Presley (at the front), with his father, Fred, at the rear.

know what happened to the insects that must have been in the straw but the mattresses were very comfortable.

Patricia White

Posh Accommodation

We Horns were a big family. Sophie is now Mrs Taylor and Rosie married a Hallam. We came from Dockhead in Bermondsey. We all went picking except Dad and Uncle Dickie who were pole pullers. We went down in April to string the bines. There were two ways, 'umbrella' and 'wire'. With the umbrella method we had to step over the wires to get at the hops because they were strung both ways. We children went to Mereworth village school, Wateringbury. One year, Iris had her own half bin as she was saving up for a bike. She saved £6 so had £1 over. There was a German POW camp at Mereworth Castle and the Germans knocked us two huts together to make one large one, put in new windows and made a proper bed with drapes like a real four-poster. We cooked on Calor gas but before that it was a Valor paraffin heater. Granny Taylor died at ninety-six and she was hopping till the last. Our family and the Sullivans always went hopping together. There were five of them: Don, Alice and three children. We went to Mereworth to the cinema or a dance seven miles away. The men hid in the hedge while we girls tried to get a lift in the army lorries. When they stopped for us the men jumped in too. Mrs Smart in the village cooked our Sunday roasts for us for a shilling. We

box were put on the back of a lorry which took us to New Cross station. We walked from Paddock Wood to Tent Common, pushing the box. There was a row of corrugated iron huts which was our home for the next few weeks. Inside each was a bed made of a wooden frame with a base of slats. We all slept in this: mother, father, brothers and sisters. It was quite cramped as you can imagine because I had a lot of brothers and sisters. It was very noisy if it rained because the rain beat down on the corrugated iron. Also there was no privacy as we could hear every word said by the families in the adjoining huts. The farmer filled one of the huts with straw which we used to fill pillowcases and mattress cases to sleep on. I don't

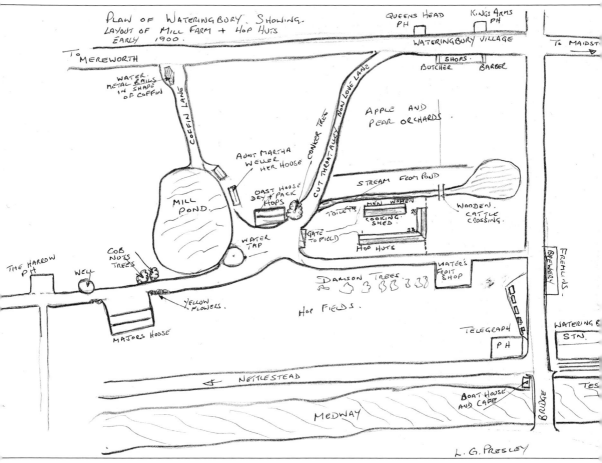

The following labels appear on the plan:

PLAN OF WATERINGBURY. SHOWING. LAYOUT OF MILL FARM + HOP HUTS EARLY 1900.

QUEENS HEAD PH · KINGS ARMS PH
WATERINGBURY VILLAGE
TO MAIDST
TO MEREWORTH
SHOPS · BUTCHER · BARBER
WATER. METAL RAILS IN SHAPE OF COFFIN
COFFIN LANE
CONKER TREE
NEW LOVE LANE
APPLE AND PEAR ORCHARDS.
AUNT MARTHA WELLER HER HOUSE
CUT THROAT ALLEY
STREAM FROM POND
OAST HOUSE DRY + PACK HOPS
MILL POND.
TOILETS · MEN WOMEN · COOKING SHED.
WOODEN CATTLE CROSSING.
WATER TAP
GATE TO FIELD
HOP HUTS
THE HARROW PIT
WELL
COB NUTS TREES
WATER'S FRUIT SHOP
DAMSON TREES
BREWERY
FREELANDS
WATERING B STN.
YELLOW FLOWERS.
HOP FIELDS.
MAJORS HOUSE
TELEGRAPH PH
WATERING B
NETTLESTEAD
BOAT HOUSE AND CAFE
BRIDGE
TES
MEDWAY
L. G. PRESLEY

A plan of Mill Farm, Wateringbury, drawn from memory by Leslie Presley.

put the meat in the middle and potatoes all round.

Iris Hans

Treat

I didn't like the toilets. They were only holes in the ground and were a long walk from the huts. I remember the smell of hops on my hands. When we stopped for our mid-morning break we couldn't wash before eating so everything tasted of hops. We called the hop-pickers' shop 'The Shack'. I

enjoyed their seafood, shrimps and winkles as a treat. We ate them as they were.

Marian Richardson

Faggot Fires

I lived in Greenwich but was born in Catford. My grandmother lived at Dean's Farm in Cuxton; Lord Darnley owned it. Granddad and my uncle George were both wagoners and I went there for my holidays. We cooked on faggot fires and wrapped herrings in wet

33

Susie Gower gathering the firewood.

newspapers to cook in the hot ashes. By the time the paper burnt away the herrings were cooked. When Uncle George got married grandma went to live at Bush, as uncle's house was a wagoner's tied cottage. One of Lord Darnley's gamekeepers lived next door to him.

Richard Phillips

Horse Stall Bed

I was born near Fordcombe, right among the hop-picking gardens. I picked every year as a child and as I got older I was shown how to do the twilling as well, which was trailing young plants round the strings. When I was in my teens we moved out of the country to Croydon, which I hated. I've always loved the countryside which is why I came to live in Biggin Hill. We went picking at all different places round the county. It was like a holiday. People came from London to some gardens. Where I went the farmers gave us a barn or horse stall to stay in. We slept on straw. It was great. Out in the hop gardens we had sack bins to pick into and a straight way between the tall poles. I used to think it was like ships on the sea. We were up early mornings, wet or dry, and stood by the bins, picking. No leaves; the hops had to be clean. About mid-morning the Measurer came to collect the hops with his bushel basket. For every bushel you were paid sixpence or a shilling according to the size of the hops. Some were small, others, Lady Hops, were double the size. What you earned was marked on your card, which you kept safe. For lunch time we stuck a sprog in the ground each side, a stick across the top and hung our billycan over the fire to boil water for mugs of tea. It tasted good. We stopped at five. It was a long day. We made another fire back at the huts and cooked dinner which was usually corned beef, dumpling stew, bacon pudding or pies, or we'd go in the oast house where they dried the hops and sit round the fire with sausages and baked potatoes. I loved it. As a matter of fact I still make those meals.

Elizabeth Webb

Nannies

My mother and the whole family began picking hops in the 1920s on Mr Goodwin's farm, Wateringbury. She went fruit picking first, then hops, then vegetable picking. We travelled down by lorry, slept on straw mattresses and everyone had to pick as many hops as they could. It was money for Christmas or new clothes. At the end of the day we all walked back to the huts. Our evening meal was nearly always corned beef stew, vegetables and potatoes baked in the fire. There was a big roadside barn, Nannies, where we could buy a pint. Sometimes at the weekend we went down to the Queen's Head. We all took a lot of pride in our work and have a lot of happy memories.

W.C. Langley

Moustache Cup

We usually went down with two other families in a big lorry and we went by the Hoppers' Special twice. Once I started working I didn't go, but mum went down a couple of weeks early before hop-picking started on the first of September. Our farmer wasn't too keen on this and one year he refused permission to go on the farm before time so they either had to look for accommodation at the workhouse or stay with people they knew among the home workers. We had a good friend, Aunt Esther, who lived in the village. She looked like a gipsy. She took the family in. My brother Walter was thirteen. They ran out of cups so he was given a moustache cup to drink from. Wallie died last year of a heart attack. We basket pickers were paid by the basket and got more than bin pickers. We were paid 1s 9d or 2s for a basket,

Jim Wood's mother preparing an alfresco dinner.

which held six bushels of hops. We had to pull our own bines and worked down the alleys where the hops were bigger.

Charles Whittle and Joan Kendrick

Travel by Steam Tractor

My husband and his brother Fred first went to the hopfields in 1932 with a neighbour on a steam tractor. All the hot smuts from the smoke stack were pouring out onto their faces and it was a really uncomfortable ride, especially as they had to keep stopping to fill the boiler up with water.

Mrs V. Hammond

Susie Gower in the tub at Chambers Farm, Benover, Yalding in 1956.

By the Bushel

We lived at Otford and my family were pickers for many years. I picked from about 1922 until I left the village at fourteen. We'd walk from home and pick in the local fields, starting work at around 8 a.m. and working a good many hours. I went with my mother and two sisters. Dad worked at the kilns. We were given a bin or half bin and picked hops into them. Small hops took longer to fill up the basket. Then they were measured by the bushel and emptied into pokes. The number of bushels we'd picked was written by a man who kept our records in a book. The rate we got paid varied, but was about 1s for four or five bushels. Some farms nearby picked into large baskets, not bins. We usually took our own food; bread, cheese and bottles of water. If we picked in hop gardens near Otford station we got our drinking water from St Thomas à Becket's well. Twice a week a man came to sell pies and cold fish. The sweetman came another day with large humbugs on his tray and sometimes toffee apples. Mother seldom bought anything because we were too poor. Villagers weren't keen on London pickers because they called strikes, not being satisfied with the wages.

Mrs B. Easdown

Dad was a Pole Puller

Our beds were made up of straw mattresses and you could feel the mice moving about in them sometimes. We had paraffin lamps for lighting and the paraffin van came round each week. Even now the smell of paraffin takes me back. My mother cooked for the family in a brick-built cook house and we collected faggots from the farmer's yard for fuel. Our toilets were chemical and they had to be emptied periodically. I hated picking hops because they made your hands black and everything you ate smelled of them.

Patricia Facer

I Hated It

I went hop-picking with a school friend and her mother and brother when I was about seven. We all slept in a

Gwen and Rita having a quick wash.

double bed that seemed to fill the whole shack. I was covered with insect bites. The food was awful and we had to get up every morning and pick hops. They smelt terrible and stained my hands. I'm telling you this because every account I've ever heard tells of the fun and lovely weather that was enjoyed by everyone but it rained when I went. I hated it. I remember just wanting it to end.

Mrs M. Showell

We Were Up at Six

We lived in Swanscombe. Every morning were off at six to the hop gardens in Betsham. Mother had drinks, sandwiches and usually some sweets in a big pram. I remember the sun shining and the birds singing. We were really happy. Mother shared a bin and my brother and I picked into an upturned umbrella. About twelve o'clock a man walked down the aisles carrying a big tray strapped round his neck. We'd hear him call 'Chelsea buns, all hot!' and were up and running, mother behind us with the money. At four o'clock the field foreman called 'Pull last bines' and we'd all go home, but we'd be up at six again the next morning.

Joyce E. Fuller

Old Friends

I was nine when Gran took her young family and me hopping in 1930. It was like a holiday for me but it earned Gran a bit of extra cash to help out with the family budget in hard times. Granddad got the old kitchen table and nailed boards round the legs to make a large

Terry Sullivan with Mark Streak, David Wheeler, and a hopper's toilet, its seat propped up outside.

crate to hold all our things. We had food, crockery, pots, pans, washbowl, toilet things, changes of clothing, mattress covers for filling with straw, blankets, candles and a piece of carpet for the hut floor. Some of our neighbours clubbed together to hire the local greengrocer's lorry. He took us from Bostall Heath to Selling but Londoners mostly travelled by Southern Railway's special trains. At the hopfields we met up with many old friends from previous years. We were allocated a hut before going to collect faggot bundles from the farmer while Gran organized the cooking outside the hut. She built a fireplace of bricks and clods of earth, put steel rods across and a tripod with a hook to hang the pots on. We made up a large bed for the children. We slept head to tail, boys one end, girls the other. Gran had a small bed. We'd stand the table up ready for our first meal and then it was bedtime.

W. Marshall

Caravan

We, that's my mother, grandmother, aunts and all the family, went picking at Merriwall Farm at Staplehurst. I have aunts who live at Slade Green and Belvedere. We used to take our own caravan and stay a few

Mrs Jellings preparing dinner on the pram outside her hut.

months. My Gran gave birth to her last child there.

Mrs B. McAlavey

Woodfalls Farm, Yalding

We were met at Paddock Wood station by huge farm carts and taken to Woodfalls Farm, Yalding. There were about forty to fifty huts on our farm. Ours was very basic. It had a tin roof, a dirt floor and no lighting or water laid on. We collected faggots and laid these on the hut floor for a bed. We put blankets and whatever bed-linen we'd brought with us on top of these. We'd cook on camp-fires outside the huts or on a wood stove in the cookhouse. Meals were very basic and the sanitary conditions were 'quaint' to say the least: the farmer dug a hole some distance from the huts and put a thunder-box over it. Every now and again they threw disinfectant powder into the hole. You can imagine what it was like come August! We children were forbidden to use this toilet and most of us went into the fields, so you had to be very careful where you were stepping.

John Meinke

Ten in a Bed

Our accommodation was a row of corrugated iron huts for sixteen families. There was no heating and the bed was communal – and I mean communal! The farmer supplied faggots and straw for the bedding. At the end of

Shirley Whiting's mother at the cooking fire.

doors. Another row of huts in front of ours were the cookhouses. Inside each was a fireplace made of bricks. Across the middle was an iron bar with several hooks holding cauldrons like the witches'. All the vegetables and meat were cooked in these pots when we lit a fire underneath. Our fuel was faggots which were twigs in bundles, rather like witches' broomsticks. Faggots were stored in a nearby field and families had to collect what they needed each day.

Patricia White

Paraffin Drink

We went down before hopping started and papered out our hut. It was hard work. Auntie Annie Edwards was very house proud and used to take the best of her sheets, china and curtains down with her and she wasn't shy of showing it off. One year she went by train and had everything in her hopping box. Coming home she got to New Cross and found someone who'd got out at Grove Park had taken her box and left a box of rags. When I was six I asked for a drink of water. Mum said the bottle was in the bin, but so was a bottle of paraffin for the lamp and I'd taken a good swig of it before Mum saw what I was doing. The farmer took me up to the Cottage Hospital on his horse. I don't think they did much to me, although I was put into Marden Hospital. The Salvation Army were very good. They came onto the garden and we kids gathered round and sang songs. They'd give us each a big arrowroot biscuit.

the season there would be hardly any left. It was used to start the fires with in the mornings. The food we ate off those open fires tasted wonderful. We were all up at six thirty. The dew had seeped into the hut so everything was damp and cold. Nobody wanted to get out of bed. Mother had been outside and lit the fire using faggots. We'd drink a mug of tea made with condensed milk, which was really welcome.

T.E. Fielder

Morning Call

We were woken at six each morning by banging on the hut

The farmer's boy sold us rabbits for about fourpence and pot herbs for twopence. The stew would be cooking all day. No-one ever touched anyone else's grub. On weekdays after coming back from the hop garden we'd have a wash, eat, then fall into bed.

Joan James

Hut Four

We went to Hall's Farm in Marden. Bert from the farm collected us from the station with his wagon and horse and took everyone back to their huts. Our wellies were never new; Mum bought them second-hand in East Lane market, Walworth. We had a bale of straw for each hut and we'd fill the mattress ticks with it, then make the bed with wooden planks the width of the hut. We always had hut no. 4, the Westcotts had no. 3 and the Jameses no. 2. Mrs James always had a new baby when she arrived. Some people put wallpaper up, which wasn't easy in a galvanized tin hut with no windows. We started work early. The bines were wet and some had long, green caterpillars on, which I didn't like. We'd have a mug of tea and some sandwiches for lunch with our fingers all green or black from the picking.

Joan Lewer (née Thirkell)

Moles

At Wateringbury we lived in old tin huts with earth floors. When it rained the moles used to dig their way up through the hut floor while we were sitting there and we threw things at them to get them out.

Ernest and Shirley Whiting (née Southgate)

Dolly and George Thirkell.

CHAPTER 3

Working the Hop Gardens

Bob Orris's family hop-picking.

Doreen Dillon with her grandmother and mother in 1953.

Dock Leaves

They woke us at six o'clock in the mornings for a seven o'clock start to the picking. The farmer boiled hot water in the boiler house, first thing, so we children had the job of taking the teapot round to make tea. There was only one cold tap to serve everyone there. We queued for every drop before carrying it back to the huts. We walked to the hop fields through mist, rain, frost and dew; it was only if there was a deluge we were allowed to start late. The farmer checked the huts each day to make sure no-one stayed behind as everyone had to work and we were there to earn money. The bines were wet and cold. We got showered every time we pulled a bine down so, for the first part of the day, wore macs and sacks until the day warmed up. My sister remembers fires being built in the fields because of frost, but I remember all the days as being sunny. We got an hour for our lunch, eaten on the spot, usually sandwiches held in a bit of newspaper as our hands were all black and tasted bitter. There wasn't anywhere to wash and if you wanted to go to the toilet you found a hedge and a dock leaf. We worked until five or six in the evening according to the needs of the oast houses and how much we'd picked.

Joan Jeffery

Cleanest Picker

My sister only did one basket a day but there were no leaves in it, so mum saved hers to put in the tally basket last. One day Mr Bones and the foreman, Leo, came to collect the tally. Mum's tally was ready but Leo asked for a 'few more on top, please Mrs Wallace.' She nearly died when he said 'Not so

43

Billy Wright pulling bines.

many leaves.' She'd used her own basket by mistake. This became a family joke.

Doreen Dillon (née Wallace)

Hard Work

We had poles for hop training in those days. In February the dirt was moved from the plant and the men sliced into the heel of the hop to start the new shoots growing. You'd get many bines from one heel and we trained them up the poles about April. We'd put nine bines up the pole and tie them with rushes picked and dried the summer before, making sure it was tied well. You train hops the opposite way to runner beans; anti-clockwise. There could be thousands of bines to do and you had to do 900 of them tied three times before you got thirty shillings. I often cried. Having two kids to pull in the pram and what with feeding them and everything I'd be well behind the other women. Their children were grown-up or at school. I'd be up at 4 a.m. to catch up. My husband looked after the boys while I was gone. He'd to be at work at seven so I took our clock to be home to wake him for work. We'd have to put wires and strings up for the bines to climb. We'd both be on stilts: my husband pulled the wire to the next pole and I handed him the hooks to fix the wires, then fed the wire to him to fasten it. The men did the stringing. The ball was in an apron pocket and they walked along, hooking it up to the wire and down to the heel using a special hook, four strings to a wire.

Alice Heskett

Stool, Apron, Basket

We usually arrived on a Saturday and started picking on the Monday. The first day of work we were allocated a row where we had to pick. We'd be armed with a stool, an apron and a basket and we'd pull down a vine and start picking. We kids were told how many baskets we had to pick before we could go off to play. After, we'd fetch more water and go off to the farm shop for bread and milk.

C. Mortimore

Florence Wright, with Roger, Karen and Della.

Nicotine

Before the hop-picking season was the growing part. When the hops were nearly ready and had grown up the wires they had to be sprayed with what I think was a nicotine solution.

Mr E. Grimwood

Smells and Stains

There was a terrible smell when the weather was dry, especially from the oast houses because of the sulphur burning. But I loved going upstairs to the drying room. The hops smelled sweet in wet weather but then they were terrible to pick because they weighed heavy and squashed easily so it was harder to fill the bins. At least your hands stayed clean! They got very sore and cracked and badly stained during the weeks.

Enid Styles

Alarm Call

Our peace was shattered first thing in the morning by a bugle blown about half past seven. We had to be in the fields by eight ready to start picking. I'm sure no-one who's ever picked hops forgets the experience of vines covered in early morning dew. When you pulled the vine you got soaked. Hops and vines are very abrasive and in no time our legs and arms were covered in scratches. As you picked your hands were stained black from the hop juice which was hard to wash off, so when you ate the food tasted foul unless you held it in paper. As a kid I sat by the bin, picking

45

Becky Napp, Eileen and Ann Reed, Ted Newton and his son, and May Newton.

into an upturned umbrella because I was too small to reach the bin. If I wandered off I got a bashing.

John Meinke

Bowls for the Children

When we were in the fields in the day, our mums picked in bins and us children picked in bowls and tipped them into the bins before the binmen came around.

Mrs V. Vaughan

Drowned in Dew

When we got to the hop field we were expected to start picking straight away. The bines were all wet and dewy and when you pulled the first bines you got drowned. The man who set the bins on was Mr Castle. He was from West Ham and was also a pole puller. We got paid a shilling for every five bushels we picked.

Arthur Smith

Nature's Swings

I loved swinging on the bines to get them down. Some were hard and some easy so we used to fall over or climb on the bins to get them down. The poleman came along with his hook if they were hard.

Enid Styles

Mrs Dalton's family.

Twilling

We women trained the hops. It took the skin off the back of your fingers and cut your palms. Hard work. We put three bines on one string and two bines up the next, close to the string so the tractor's shimmer wouldn't plough them out or you'd have no hops. As they grew, we had to strip the bottoms of the bines or they grew like bushes and no air got to them. If the bines hadn't grown over the tops of the wires by 21 June (midsummer) it would be a late hop-picking. If hops got red blight we washed our wellies and clothes with disinfectant and changed them in the field tent so the blight wouldn't spread or it would kill all the hops. We'd string a diseased bine off and burn it.

Alice Heskett

Versatile

I was seventeen in 1944 and worked as a telephone engineer for the Post Office. We got annual holidays but the Government encouraged us to stay at home because of transport difficulties. Most workers were traditional hopping families and not many of us smaller volunteer groups. We lived in basic bell tents with a rubber sheet on the grass and minimum bedding. Meals were provided by a sort of field kitchen. Pickers were there only to pick hops and we were there to make sure they could, unhindered, by helping the full-time farmworkers. I spent my time on a variety of jobs, assisting the tallyman, measuring the pickers' hops into bushel baskets and entering the amount on the tallysheet so they'd be credited with the correct amount for payment. We emptied the baskets into hessian sacks.

Mr Homewood taking the tally in 1961.

Pot's Boiling!

We all made up our huts with curtains and wallpaper. They were lovely. We had to be up at six every morning and a man used to shout 'Pot's boiling!' for hot water. The Measurer came round, tallying hops from the bins. My brother and I had half a bin. We got a shilling for every five bushels picked. No leaves or bits of vines to go in them. We bought our veg locally and the butcher came once a week so we had meat, but up the fields at dinner time we had bread and apple sandwiches. We picked until the heavy machines came. We got our fingers stained but we didn't care. We loved it.

Mrs M. Sanders

Hard Work

That first day was all rush so we'd be in the field by seven or half past because we wanted to look at the hops to see how large they were. We'd try to choose a bin next to a friend but not on the edges of the field where the hops were poorer. It was best to make for the centre aisles; the hops were denser there. With luck we'd get a lift in the binman's lorry, else we'd have to walk and that could take half an hour depending on how far off the field was we had to pick. All the fields were different acreages. We'd have to be on the field and started picking by eight at the latest and go through to four or five-thirty except for lunch break. We were allowed tea breaks when we liked. The kids had to pick all morning but they could play in the afternoons. They

When they were full we put them into a big, two-wheeled, horse-drawn cart. It rained most of the time so it was muddy. Apart from the discomfort the rain made the hops and sacks much heavier and harder to lift onto the cart and then even more difficult to move the cart. We eventually coaxed the horse to move it to the oast house. Being quite tall I sometimes got the job of bineman, patrolling the picking area armed with a special pole to pull down the broken bines for the pickers to finish off. My first morning in the wet tent I picked up my enamel plate to get my porridge and found it covered with slugs that had crept in during the night.

John A. Graham

helped when the tallyman came round, taking the leaves out of the bins. We didn't need any tools, although some wore rubber gloves so their hands weren't stained. We were allocated a row and a bin and they were all lined up the first day for us. We didn't have teams. You kept the same bin all the time and it had your number on it. We put small tots into one end of our bin to play or the big ones looked after them or they sat in their pram. The binman helped ladies on their own to move their bins, which were heavy and awkward. You could pick about eight bushels a day. I think we got about two shillings a bushel. Some had five or six bins for a family, but we were only mum and one child so we had just one.

Enid Styles

Home Dwellers

We lived at Dunton Green so the hopfields were only a few miles away at Shoreham and Otford. I remember being dragged out of bed while it was still dark. We were given a quick breakfast before being rushed round to the pick-up point. We were taken by lorry and we children used to kneel or stand, looking over the tailboard. It always seemed either chilly or foggy first thing but the sun warmed up the fields during the day. From a child's point of view the hopfields were a natural playing area. It was fantastic for making dens around the bins and playing hide and seek. It wasn't all play though. Most of the children helped to pick hops for as long as they could then were given time to go off and stretch

George Thirkell at the bins.

their limbs. There seemed to be very few men, if any, at the picking but men would come round the bins to empty and weigh the hops. There were lovely smells: the aroma of the hops mingled with the smell of the sacking the bins were made of. I can remember that smell even now. All the women seemed to pick at a furious rate, so it wasn't a leisurely occupation. There seemed to be an air of competition around as if there was a race on to get to the end of the row first. It was certainly an incentive to the children to give a hand. I don't remember any Londoners or living quarters so maybe the farmer only used local labour. It seemed a long day, out in the open air all the time. We were extremely grubby at the end of it,

Lily and Doreen Whiting.

but I seem to remember sleeping very soundly at night. We didn't go away on holidays then, but it was a great adventure to us children. They were happy, carefree days.

Verna Hunt

Soaked

There were my sisters Ruby, Cissie and Lil, my mum (Flo Ritchie) and me. Ruby was an excellent picker. She got every green leaf out of the basket. I didn't like hop-picking. When you'd finished picking your bine you had to call the pole-puller to get you another. He'd use a long pole to catch the bine at the top of the strings. If it had been raining we'd all holler because we'd be soaked with water. One day we went to the oast house to be shown what they did with the hops to make beer. They didn't smell nice. I worked as a shelf filler for eighteen years at Somerfields in Belvedere – it's Wallis's now.

Edie V. Mortimer (née Ritchie)

Trainspotting

There was a high demand for bins and sometimes only half a bin was available or shared with another family. We travelled daily to the hop gardens from Northfleet. The farmer only sent for London pickers if there was no local labour as it was expensive having to provide accommodation which had to be maintained. I spent a lot of time trainspotting at the nearby station and remember seeing Schools Class and Battle of Britain Class engines. Where your bin was affected your earnings: at the end of the line the hops were often smaller because the fertilizer hadn't reached there, so there were fewer hops on the bine. Some people tried to influence the farmer into giving them a better place, saying they were old, infirm or needed to be near the tap. A big, strong woman did the measuring at Woodwards Farm. She tipped the bushel baskets into a long sack, a poke. When it was full, 'Poke!' she'd say, and the men behind her lifted it onto the horse-drawn wagon. They reckoned a gent's umbrella held a bushel and people picked into upturned umbrellas if they couldn't get to a bin.

Mike Pullen

Toilet Made for Two

I was eleven (in 1946) and came to live with my Aunt May and Uncle Albert when my parents were killed in the war. She'd been a hop picker for years so I went with her on her annual jaunt. Aunt May came from East Ham and Plaistow and still had many friends there. We went with them to Boughton for the picking. Aunt and uncle took me down a couple of weeks before the season started, sometimes by train and sometimes on the motorbike, did any repairs and gave the hut a good clean. Although we weren't allowed to, Aunt kept the hut padlocked so no-one else could use it. On our side of the field were all our friends, and people from other areas on the other side. We were always the last family to be picked up, living in Welling. The others joked about Aunt's mattress which was hard to fit into the already crowded lorry. We worried we wouldn't make it up Star Hill in Rochester but the lorry chugged its way to the top and we'd all cheer. The toilets were just wooden huts in the middle of the field. They covered a deep hole and there were two round holes in the seat so two people could go at the same time. It was great picking when the sun was shining. We got lovely tans and looked like gipsies with our turbans on. We shared our fire with other families and there was a sort of tripod over it to hang the kettles.

Shirley Thorne

Regular Job

I married a hop farmworker on Grange Farm, Tonbridge. When I first started I'd never known farm work but suppose I accepted it. It was very hard. I went to work at 8 a.m., came home at midday, lit a kitchen fire to fry bacon and bubble and squeak, then was out in the fields again at 1 p.m. We came home at 5 p.m. and I'd cook a proper dinner for us – you always cooked enough so you could have enough left over for lunch the next day. Then there was the children's washing to do and get them to bed. On Saturdays I walked five miles from Grange Farm to Tonbridge for the shopping. I baked twice every week, made jam, bottled all my fruit, everything. I worked on the farm haystacking, very heavy work, and learned to milk cows and thresh – the filthiest job on earth – the machine would throw all the dust and stuff at you.

Alicia Heskitt

Tally Sticks

I was born in Chislet in Kent and my grandparents brought me up. They worked for a local farmer and the wives worked three weeks in his hop garden every September. That was our summer holiday. We got up very early by candlelight, then walked two miles across fields, getting wet from the early morning dew, ready to start picking by 7 a.m. We picked into bushel baskets or tin baths; anything that would hold hops. I sat on an upended wooden box so I could tuck my feet in out of the

Nina Ford, Mary Stamp, Ethel, Tom Gower and Terence Stamp (who later became an actor) in 1943.

cold. A farmworker cut us down some hop bines with a knife on a long handle then he'd call 'All to work,' and we'd start picking. The baskets held one bushel. When it was full we called for the tally. We tipped it into a five bushel basket and the tallyman came down the alley and matched his and your tallysticks, marking a line across the two with a file. We treated every hop as if it was gold dust, picking up every one that fell on the ground. If they were a nice size it only took about forty-five minutes to pick a bushel; if they were small it took a lot longer. At about 6 p.m. if the oast was full or if we'd not picked enough hops to make up five bushels, we were given tokens to use the next day. We walked back home to our farm cottage and prepared dinner, then packed up food for taking to the hop garden the next day. The farmer treated us all to big pears out of a basket. We spent all Saturdays and Sundays cooking and baking for the week, doing the washing and cleaning our boots ready for Monday morning again. It was hard work and very tiring. I got sore arms from all the scratches from the rough hop bines. I got my reward at the end of the season: a pair of new lace-up boots for the winter and a pair of thick, black stockings.

Lily Dicken and Edith Bartlett

Clean Picking

Each family had a lane to pick and you kept to the lane until the field was finished. If it was a big field pickers began at both ends, and met in the middle, but we guarded our lane jealously if someone tried to pick off our bines. When the bine 'heads' got stuck on the wires we called a pole puller to pull it down as the best hops were on the heads. Last thing in the afternoon the tallyman came to see how many bushels we'd picked. I daren't knock the basket or the hops settled and we got paid extra for full baskets. My hops went in last as I picked clean (no leaves). Baskets with leaves in got marked down on the tally.

Shirley Thorne

Apple Scramble

The hop fields were quite near the huts and we were usually at the bins by eight o'clock. The bins were about seven feet long by three feet wide and three feet deep and held twenty bushels. Binmen came to each bin and measured the amount of bushels using bushel baskets. Payment varied each year. One season it was three bushels for a shilling and another it would be five. Not a lot! Usually our season lasted three to four weeks. At weekends older members of the family came down to visit, either by train or cycle. They'd bring changes of clothing and some of the goodies we were missing, like cold roast lamb or beef. It certainly made a change. On a Sunday the farmer sent a lorry down to our huts which were in a large field, and they'd throw apples over the back of the lorry for an Apple Scramble. All the children and grown-ups picked up as many as they could carry. It was great fun.

T.E. Fielder

Comradeship

There was great comradeship in the fields, what with singing all the songs and the wise-cracking. My gran would call out, 'Come on! Wire in. Get picking.' Lunch was sandwiches, fruit and a drink, eaten with stained fingers. Our favourite words of the day were 'Pull no more bines', the call to stop work. On the last day excitement was always high. We looked forward to the fun and games after the last bine had been pulled. People collected elderberries ready to blacken the face of the person who pulled the last bine, then it was a free-for-all with everyone trying to blacken everyone else's face they could reach. It was a last chance to have a ride on the hop cart. When pay day came it meant a trip into town and we were bought a present for all our hard work. Then it was time to pack up again and wait for the lorry to take us home and back to school.

C. Mortimore

Dad's Lift

The men came down Saturday afternoons. They had to work in the mornings if they had a job. My dad waited by the side of the main road at New Cross to be picked up by one of the vans that had been up to Covent Garden and he'd get down to Marden for the price of a drink to the driver.

Joan James

Counting the bushels.

Pains and Stains

We had two huts. Our bed was made of wood and wire mesh with bundles of straw laid across it for our mattresses. There was a cookhouse for our twelve huts. We had porridge for breakfast and stew for the evening meal or maybe fish and chips from the mobile van. We bought meat and pork pies, bread and sausages at the local shop for lunches and made a fire in the fields to cook our sausages. If we wanted a cup of tea we'd boil cans of water, put in some tea leaves, sugar and milk, stir it, then pour it into mugs. It wasn't bad, ha! We were up by 5.30 a.m. and started work at 6.30. The farmer told us what field to pick and we'd finish about 5.30 p.m. in all weathers. We walked to the nearest fields but if it was quite a way the farm labourers used to take us by tractor and we'd sit on the trailer. My hands got very sore pulling bines and the hops stained them black so I'd have to scrub hard to get them clean – I hated that.

Brian Smith

Happy Lot

I went with my two boys and baby daughter when she was only $4\frac{1}{2}$ months old and the boys were eight and five. People used to look down on hop-pickers as being dirty but, believe me, you could be just as clean if you've been used to it. We went picking at Cowden near Penshurst with our neighbours. We were all good friends. They were a happy lot and we all got on well together. It was early when we went in the field and when the sun came out it

54

was lovely. We could smell the hops as we picked them. We stopped for dinner at one o'clock but our fingers were black with the hops so we had to hold a piece of paper round the sandwiches.

Mrs Q. Moody

Gran Died

We kids loved hopping but we had to pick all morning before going off to play. We were at Five Oar Green, Pembury. I was only ten years old but I had my own half bin and by the end of the season I'd earned £2 10s and that was a lot of money for those days. The farmer gave me half a crown for myself. 1936 was a wet year and we couldn't get up the gardens (always called gardens, not fields) for days. Mum went down with a chest illness and had to be taken home. I was left with Gran and four days later Gran died. The farmer was very good to me. We got her body home to be buried in the Brockley family grave and we had to pay every place her body rested, but we whipped round and got her home.

Joan James

Picking

Hop vines were grown up wires, rather like runner beans are grown. A member of the family pulled down a vine and hung it across a bin so we could all pick the hops and drop them in the bin. We weren't allowed to put in any leaves or bits of vine as this was 'dirty picking'. When the first vine was

Tom Jnr, Tom Snr and Ethel Chandler with the rest of the family on May's Farm.

pulled we'd hope it hadn't rained in the night as the water showered down on us and we'd get soaked. After we'd picked for a few hours they'd shout 'pick no more bines.' All the picking had to stop while the Measurers came round to judge how many hops had been picked by each family. A note was made of their tally in the family's tally book which they kept. At the end of the picking season our book was taken to the farm manager who'd pay us the going rate for each bushel. I remember mother telling me they went picking to have a holiday, but in truth it was to earn the money. At the end of the season the farmer would let us use a hut to store pieces of carpet, saucepans or anything we wanted to store for the following year's hop-picking session.

Patricia White

May's Farm Tragedy

I was born in Virginia Road, Dockhead. Mother took me down hopping the same year. I never missed until they brought in the machines. We lived in a cul-de-sac and nearly everyone went hopping except Lily Rose's family. She was my best friend and her father was terribly strict. There were the Gosbys, the Hollands, Rileys and MacDonalds; like one big family. Children today don't know what it's like to live. One year my dad couldn't get any work on the docks so we went to Mr Bett's in March, weeding and pea-picking, then across to Mr May's for the hops and back to Mr Betts for the potatoes.

We didn't get home until October and did that for two years. It was lovely. At Mr May's if you had a fire after 10 a.m. you didn't get a faggot the next day. One year my aunt came down. We'll never forget her husband, Harry Bush, sitting there with a penknife cutting the hops off one at a time. A man boiled water up for the pickers in a big copper every morning and night. One couple went out for a drink and put the three children to bed. They made a terrible mistake: instead of putting a twig in the hasp they put the padlock on. I don't know if they knocked a candle over but no-one could get to the children and they burnt to death.

Ethel Chandler (née Ash)

The Woods family.

CHAPTER 4

Love Bines Strongly

'Togetherness'.

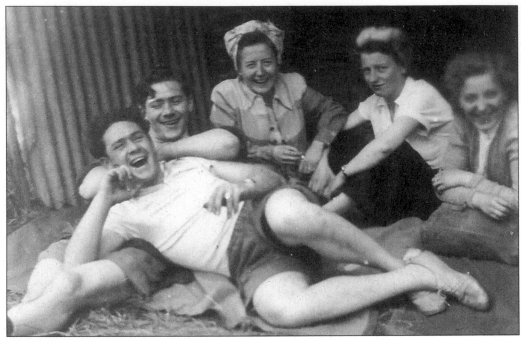

'Having a laugh'. Doris is at the far right. The others are, from left to right: Billy and Albert Turner; Albert's wife, Sheila; Doris's sister, Rose. The picture was taken by Doris's husband, David, at their hut.

Doris and David

We'd spend six weeks of hopping holidays down at Goudhurst. My parents were Lillian and George French. I was the youngest of nine children. We came from Stepney Green. My dad's sister, Daisy, and her little son Michael came with us to Mr Barden's Farm at Yalding although our family had been to Cranbrook, Paddock Wood and Wateringbury in other years. I met my husband-to-be, David Turner, down there in 1947 and we were great pals. He and his family went to the same farm as us the next year too and we were married.

Doris Turner

Faversham Explosion

I was nine when dad died in the 1916 Faversham munitions disaster, so we lived with my aunt and uncle at Boughton. He worked in the oasts. Dad is on the memorial in Boughton churchyard. Mum cleaned the church and hall. I worked with hops from early spring until they peaked. There were four Boughton pubs. Hoppers' husbands came down at weekends so the pubs were always full. We made good money collecting empties: 2d on a pint bottle and 4d on a quart so were always scheming; more bottles meant more money. We courted the London girls who came hopping and once I went up to Whitechapel when one of them invited me to visit her.

Jim Gilbert

Mr and Mrs Rule outside their hut.

Mrs Fagan's mum and dad.

Local Lads

Local village lads came to the site most evenings to court the teenage girls but my aunt never allowed me to go off with them. Most evenings we sat round the fire telling stories and singing. It was great.

Shirley Thorn

Hop Baby

We used to take the caravan down to Merriworth Farm, Staplehurst, and stay at the hop farm a few months at a time. My gran gave birth to her last child down there. My grandparents were in a book about gipsy people. I'm a great grandmother now.

Mrs B. McAlavey

Family Friends

Both our families went hop-picking with our parents. That's where we met in about 1955. We came from Limehouse and Plaistow. All the families met at London Bridge station at 4 a.m. to get the mail train. Everyone was struggling with tea-chests, bags, boxes and prams. When we got to Marden station the farmer took our heavy luggage and we walked. Then some families got together and hired a

Florence Wright's family.

Bob and Lou in 1962.

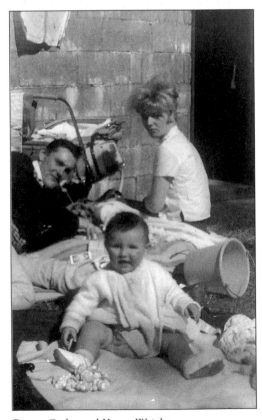

Roger, Cathy and Karen Wright.

Mrs Westcott and family.

large covered lorry to take us to our farm and that was much better.

Mr and Mrs Andrews

Hard Up and Happy

I was born in St Olave's Hospital, Rotherhithe. Dad was a stevedore at Surrey Commercial Docks and with four children, like their neighbours, we were always hard up. Our only holiday was going down 'opping. It was our parents' way of giving us a bit of country air and earning a few bob to boost the family income. I was a real 'opping kid as mother conceived me at Whitbread's Bell Common Hop Farm, Beltring.

Laura Rule

Unusual Certificate

I've taken many of my foster children to the hopfields. Some had never even had a holiday. They never knew anything like it. Della and Karen became part of my family with my own four – Billy, Catherine, George and Caroline. I loved them all. Caroline and Billy both married people from the hopfields. We looked forward to next year's hopping right from when this year's was ending. They were some of my happiest days. My husband Harry was a life-served sailor before getting a job as a guide at the National Maritime Museum. But the hopping time overran his work time so he had to go sick every September (the doctor knew why) as he had to run our boarding house the family and I went hop-picking. One year the doctor put a really odd word on

The Pilbeam family in September 1949: Mum, Irene, Alf, Alan, Derek, Pat and Jean.

the certificate and Harry wanted to know what it was. 'I don't know,' said the doctor, 'but no-one will query it, whatever.'

Mrs Florence Wright (died May 1998)

Courted

When I was about fifteen I was very grown-up for my age and the son of one of the farm hands courted me. His dad drove the shire horses and carts up to the fields. He used to pick me up and drive me to the field that was being picked and everyone else had to walk! I really felt like a queen sitting up beside him. I can laugh now when I think how I must have looked, climbing up onto the cart.

R. Vaughan

Romantic Goudhurst

My mum was Doris French, now Turner, and she met my father, David Turner, during the six weeks' hop-picking holiday at Goudhurst. They were married in December 1951 and had wanted to be married at Goudhurst but it wasn't possible because you had to have been a resident for a certain length of time. We went there several times in the seventies and both Olga (my wife's from Russia) and myself visit the area about twice a year, still.

Philip Turner

Local Lad

Our pole puller was a local lad, Ron Huckvale, who lived in the village (Swigs Hole) and worked on the farm. Our eldest sister Rene and Ron started courting. They married in 1943 and are still happily married. They live in St Paul's Cray, Orpington. If we were good when the pie man came round with sweets and lots of nice things to eat we were allowed a lollipop. The man that sold the goodies turned out to be Ron's dad although we didn't know it at the time. When Ron came to visit us mum showed him pictures of us in the fields and one of our mum making out to steal a pie and Ron said it was his dad.

Jean Pilbeam (née Bird)

'Lousy Hoppers'

When I got married I told my husband I was taking my girls down hopping. His mother said only lousy people went hopping. I daren't tell my mum – I think she'd have killed him.

Ethel Chandler

Gipsy Midwives

When we went down hopping on Overy's Farm, Paddock Wood, my wife's mother was $7\frac{1}{2}$ months pregnant. It was the first weekend of the season. She soon realized something was wrong and that the baby was coming. There were no telephones and her sister and family were in a panic. They ran to the gipsy camp for help and they delivered my wife-to-be, Jean, onto a straw mattress in a tin hop hut on 3 September. Jean and her mother were taken to Pembury Hospital. Right up until the 1950s the gipsies in the Abbey Wood and Belvedere area called at their house in Birkdale Road to ask about 'the little girl born in the hop hut in Paddock Wood.'

Bob Orris

Mrs Dalton's hop hut.

CHAPTER 5

Not All Hard Work

Florence Wright's family stops for lunch.

Two of the hayricks which were burnt down by hopping children.

Poor Dad

There were four of us girls and mum and dad and we didn't have much money so hop-picking money was spent on material mum bought at Erith to make our school clothes. Any left over was for mum to spend at Woolworth's, sixpence a time. We went to Hadlow's Farm, Paddock Wood. My mum wore a long dress down to her ankles, wellingtons and a woolly hat. We children filled pillowcases with hops, 'whether you like it or not, Edie!' The first thing Mum did was pin up all the ends of wallpaper she'd saved, all different colours, to make the hut look homely. We slept four in a bed but mum slept in a bed on her own until dad came down. He told her, 'I'm not sleeping here with the four girls in the room!' and she said, 'Don't be silly, Alf.' But she had to put up a piece of string with a blanket across it.

Edith Mortimer

Fire!

At Smugley's Farm, Goudhurst, the farmer had got all his winter hay into ricks standing on platforms. Some hoppers' children crawled underneath and set fire to it. We all formed a long line filling buckets from the tap and horse trough and passed them down along the line to each other, eventually putting out the fire. Before we'd gone hopping my mum had bought a brand new bucket but when the fire was out she found herself with this terrible old thing.

Shirley and Ernest Whiting

Faversham's Fleur de Lis Heritage Centre, housed in a fifteenth-century ex-inn, has a fascinating museum of hop-picking and rural life. (By kind permission of the Fleur de Lis Heritage Centre).

'Kidnapped by Gipsies'

My mother, her four sisters and all my cousins went picking. One year our mums arrived at the farm, unloaded all the bits and pieces like we always did and made the huts homely, while we kids ran wild in the fields. My seven-year-old cousin came running back shouting, 'Aunt Lizzy!' – that's my mum – 'the gipsies have taken Ronnie away!' He's my six-year-old brother. I can still see mum running across the fields in her old apron and wellington boots, crying her eyes out. When she got to where the gipsies were she needn't have worried. What had

happened was my brother had fallen in the water and the poor gipsies had got him out and had started to dry him and give him a hot drink. But my cousin had mistaken what they were doing and thought they were taking him away as the gipsies got the blame for things like that.

R. Vaughan

Invaded

Our huts were brick built with galvanized iron roofs. They were small, but enough for four of us. It took

66

two hours to clean out all the dust and dirt settled since last year, then we'd prepare the beds and put curtains up at the only window. We had brooms for sweeping out and an old fashioned carpet beater for the home-made rag rugs. We had a bucket in the hut at night or we'd go outside to the earth toilets all dug in a row down in the woods. All our water had to be carried from the tap in buckets. We were too tired to do anything at the end of that first day and had to be up at half-past six every morning. One night the cows got into our camp field. It caused a real rumpus in the middle of the night when they came among the huts.

Edith Styles

Wateringbury

At weekends dad cycled down to Fremlin's Mill Farm, Wateringbury, from the Elephant and Castle and back for work on Mondays. My friend, Bill Weller, often sat trimming sticks with a curved-bladed knife. These were for securing sacks at the neck. He softened up my new hob-nailed boots with axle grease to make them pliable. My favourite game was damming up the stream behind the huts, with other children helping. Once the water level was raised I sailed my boat. I still like boats. A dam under the bridge was useful when we went scrumping. We'd toss apples in the stream from the orchard and they'd be out of sight up against the dam if the bailiff came.

Mike Presley

Maud Presley, with her son, Leslie, and daughter, Doris, ready for a swim.

Hopping Home

My husband was a plumber. We were saving to buy our house in Welling. It cost £360 which was a lot in those days, so I went hopping every year as the extra money helped. We couldn't wait until September came. The kids and I spent all day in the fields in the fresh air. The kids played skipping games. One of their songs was 'My mother said – I never should – play with the gipsies in the wood; – If I did – she would say – Naughty girl to run away.'

Nell Hearson

Time for a cuddle.

Is it dinner time yet?

Lucky

One day we were snowed off. On the way back to the hut dad found an uncollected poke (a sack of six bushels of hops). He hid them in our row. The next day we had an especially good tally! Picking umbrella-strung hops meant four bines came down at once. This always seemed to happen just at knocking off time. My big cousin was home on leave from the army and came down to Homestall. He was amazed mum could cook Sunday roast on an open fire. There was always good humour from the mums around the huts while we were cooking. One day our spaniel, Peter, chased the farmer's piglets down the road. Boy, did we get told off by Mr Bones. Poor Peter had to be tied up forever. Sunday school was held on the open space in front of the huts. They gave us stamps to put in our books to show our vicar we'd attended church while away. Once dad took us all into Whitstable for Sunday lunch. I can still smell it. The end hut doubled as a first aid centre and they sold toffee apples there. On the last day after mum collected our pay she'd roll down the slippery slope from the oast house. We thought we were rich: £70 clear for our work. Dad only earned £4 a week then.

Doreen Dillon

Many hands make light work for W.C. Langley's family.

Wellies and Red Rings

We had to get up very early in the mornings. It was really cold. We walked everywhere. My legs had permanent red rings round my calves because the wellies rubbed against them. There was a gipsy encampment nearby and I'd run off to be with them. My dad was forever searching me out and bringing me back. I remember sitting outside the Admiral Jellicoe pub drinking Tizer and eating crisps while the grown-ups were inside on a Saturday night. There was my Aunt Mar, Uncle Arthur, my cousins Alex, Raymond and Philip Halsey, Aunt Minnie, Joe, Aunt Em and Uncle Nobby and cousins Gerald, Maureen and Sandra. My Dad was a pole puller and he came down at the weekends. Both my parents are dead now but my Aunt Mar in Blackfriars is alive and ninety-six years old. She can tell some tales about those days.

Patricia Facer

Children's Excuses

A fish and chip van came on the field twice a week, and Salvation Army ladies came round with hot jam puddings. We children picked into opened umbrellas and were allowed to go off to play after so many umbrellas-full. The older children used to say to me (being the youngest) 'Go and tell Gran you want to wee-wee!' so we could all get away from picking. We'd be away a long time, scrumping apples, playing round the haystacks with the dog. He always came hopping with us. When the whistle blew to finish picking it would be round to the butcher's at the back of

69

Billy and Mrs Wren with aunt Mag Ruby and her nephew at the bins.

the Bull pub to buy chops or sausages for our evening meal. We had to cook them on our own fires outside the huts and we'd all sit around them until it was really dark, singing and laughing.

Mrs N. Burney

Prickly Situation

Little Sheepers Farm was in a beautiful setting. In the morning uncle set off to pull the first bines. I was warned not to pick any hops or I'd get showered with morning dew. 'Just look after little Jimmy while we pick hops.' We were sitting beside the bin at midday with the fire going lively for tea and sandwiches when we heard a dog, as if in pain. Uncle went over to a nearby ditch to see what was happening and there was a poor dog in the bottom, its paw caught in a hedgehog's prickles. Uncle climbed down, released its paw, took his cap off and put the hedgehog in it. A few Hoppers came round to see what he'd got and one man told him cooked hedgehogs were tasty. Uncle covered the hedgehog with clay, baked it in the fire's embers then broke off the baked clay when cooked. But he didn't fancy baked hedgehog and gave it to the man who'd thought they were tasty.

Bob Richards

Boys!

There was always a queue to see the nurse on her three evenings a week. Once, my brother and cousin chased me with a frog. I fell over on a freshly tarred road and my knee swelled up so much they had to push me in a pushchair to visit the nurse.

Joyce Ashby

Well Trained

My husband took the horse and cart to Tonbridge station to pick up the Londoners with their bits of furniture and things. In Tonbridge there was a pub called The Mitre and the Londoners taught this horse to drink beer so he automatically called at The Mitre for his pint before going onto the farm.

Alice Heskitt

Uncle's Games

We went to Whitbread's farm near Tonbridge, but later to Catt's Place Farm in Paddock Wood which was smaller and cosier for picking hops and for the children to play. We usually picked in nearby fields but occasionally had a lorry trip to neighbouring fields. We all descended like cheerful locusts to strip the fields clean before returning to our huts for the night. The farmer supplied faggots for fuel and the old relatives did the cooking while we children played cricket or wandered off to explore the fields. We were city kids. Uncle John Dunster was happiest organizing races, cricket or rounders for everyone – he wouldn't let anyone rest. By bedtime we'd spent all our energy and were ready for sleep. Even now if I smell straw I think of cosy beds and us

Rosina Amis's family relax outside the pub.

all huddled together safely. Mornings at Catt's always seemed earlier than anywhere else; chilly, mists and strangely still. We cleaned the huts, carried water while the grannies cooked breakfast, then all trooped off with our food and utensils for the day as if instructed by an invisible time-keeper.

Michael J. Stark

Blacksmith

Mostly, husbands and uncles came down at weekends. I've often wondered how we accommodated them all, but we did. We only had wooden huts but some farms built brick ones.

Tom Ash and his sister outside the Bulrush.

Saturday nights were for relaxing with a few pints at the local then a sing-song round the campfire. Sunday, the visitors went back to London and Monday we were back in the fields again. Uncle Tom was a blacksmith and made Gran an oven. The others cooked on campfires. Food always tasted good, cooked on open fires. We picked blackberries for Mum to make a pudding for 'afters', made bows and arrows and went apple scrumping. A man came round selling sweets and teas during the day.

Joyce Rogers (née Tibbott)

Muddy Alley

We went to Forge Farm and Smugley Farm which had been combined, I think they were owned by Whitehead and Colman. There was a gipsy camp nearby with horse-drawn caravans. Further on along Muddy Alley was The White House which was supposed to be haunted. So was Captain Blunt's but we think that was said to keep us out of his orchard. Our post box was set in the White House wall. Hop huts were along two sides of the Common, the gipsies were on the far side and the Oxford and Cambridge Mission had a large marquee. They administered first aid and religious services and collected the names and addresses of all the children and, come Christmas time, threw a party for us all near Camberwell off the Walworth Road.

Bob Orris

Pay Varied

We did our shopping from various vans calling on our hopfield during the day. The grocers and butchers came. The bakers sold huge lumps of bread pudding. You never knew what you were going to be paid for a bushel because it depended on the size of the hops. If the summer was dry, hops were smaller, then in a wet one they might be quite big so you never knew how much you had to come until the end of the picking time.

John Meinke

Pole Hole

When I was ten we'd been up the hopfields and were just coming back to the huts past a 5ft deep sump hole for hop poles. I fell into some creosote. It was all round the sides and I couldn't get out as they sloped and the creosote was slippy. When all the fellows went by I'm standing there shouting 'Help! Help!!' but no-one saw me. My family went past, heard my voice and my brother brought a rope so I could climb out. While he was lowering it to me he fell in too, so someone else had to pull us both out. When I got to the top my mum gave me such a towsing I fell back in again. The next day we'd all been to the village and on the way home we all wanted to pee, so we used this bush. We didn't know it was a beehive and I got stung on my winkle. I got another thick ear from my mum and she said 'You needn't think I'm going to suck that out!'

Harry Henderson

Bravery

A little girl fell in the pond and was drowning. One of our chaps dived straight in although the pond was thick with weeds and he saved her, which was really courageous of him.

T.W. Ovenell

Films

We played in the fields and every week went off to the village school hall to see the latest film. It cost 6d to get in. From the week's money we'd have a 'sub' and go to the pub. I had to sit outside with the other kids with my lemonade and crisps. By the end of the season we hardly had any money left to go home with but it was nice to meet up every year and we made a lot of friends.

Brian Smith

Stamp Currency

One man had an old 'Stop me and Buy One' ice cream bike. He cycled down the fields selling the hoppers hot pies. If we didn't have enough money, he'd take postage stamps. Brenchley's put wire-netting mesh all round their counter as soon as the hop picking season started to stop light fingers. At nineteen when I was off work for the weekend I went down to see my mum for the day. From Beltring Station it was a long, miserable walk to the farm. When I got there the first thing I wanted was the toilet. They showed me where it was – just a hole in the ground.

Arthur Smith

Southfleet

I am seventy-eight years young but I remember my hopping days well. We lived in Swanscombe at the time and we'd walk across the A2 to the hopfields at Southfleet every morning. I was about five at the time.

Victor Huggett

It's ice-cream time for Sue Williams and Leslie Presley.

Home-dwellers

We lived on Murray Woods Farm at Green Street Green near Dartford. I'm nearly eighty now and remember the hop picking – it came during part of our school holidays and the farmer used to send a lorry to Dartford to get us for picking the hops.

E. Grimwood

Straw Mattresses

The lorry took us down to Hubbles Farm in Benchley. Grandmother used to take a feather mattress with us. We had tin huts full of straw so we all set about making the beds up on a platform at the back of the huts, filling covers and pillowcases. We even put lace curtains up to the door. The big box grandpa had made to carry our china and clothes in was our table, with a drop leaf to store food inside.

N. Burney

Posh Huts

We picked at Mr Bones' Homestall Farm, Faversham. In 1952 the formidable Mrs Porter from Swanscombe asked my mother if she'd like to take the place of someone who couldn't go. She said we would, although my father had reservations about the whole thing. He came from Jarrow and didn't know about hops. Mother had picked with grandmother in the early 1920s. The lorry was smelly; it was Mr Flint's coal lorry with a tarpaulin over it. It took all our equipment and about ten other families and all they needed for a four week stay. We went down by train the second year and had a new breeze-block hut. Dad put up a partition with proper double beds each side with real mattresses.

Doreen Dillon

Bull Pens

Just inside the farm were two bulls in pens, one either side of you as you walked through the gate. It was scary walking past them and we'd make sure we walked in the centre back to the fields. One night there was a loud noise and you could hear people shouting. Someone had left the gate open and the bulls had got out. Naturally we kids got the blame.

Enid Styles

Dewy Mornings

It was out of the question mother going hop-picking with a large family. Hers was a full-time job in the home. Her sister, Aunt Mary and Uncle Jim, regular pickers, offered to take me for a holiday. We were out early to catch the hoppers' train at London Bridge. Everybody was there with their hopping carts and when the gates opened there was a big rush. When we arrived at Paddock Wood we found ourselves on the road, the sun had broken through and there was a smell of hops. Aunt and the family all pushed the cart to the farm where we were staying. The first

morning was dewy, it was cold and I wasn't very happy. It was too early and I clung near the fire. After a hasty breakfast and Aunt persuaded me to go. We tried to avoid those first bines being pulled – if I didn't get out of the way in time I got showered by wet hops.

Bob Richards

Healthy

There didn't seem to be all the illnesses about that there are now, although the farmer made us go out in all weathers. You had to keep the fleas and bugs down. If any of us cried with earache our mothers filled a pillow with hops to sleep on.

Ethel Chandler

Sleeping Beauty

One teenage girl had sleeping sickness. Every now and again she was missing. We'd have to look for her and she'd be asleep in the hops. One lady always had the first hut. Her door was always open because she wasn't very well so she just sat in the doorway, enjoying the hop smells and the fresh air.

Evie Mortimer

Subbing

The farmer came round with a book to see if we wanted a sub. When we'd finished our three weeks my mum

Dolly and Bert Thirkell.

used to go and get the rest of the money still on our card. It might be ten or twelve pounds which was a lot of money then.

M. Sanders

Extra Rations

We took flasks of tea and sandwiches for lunch. They tasted wonderful. We got an extra cheese ration because we worked on the land. I was often sent into Boughton to buy groceries and loved wandering over the fields. Aunt May cooked lovely meals over the campfire: my favourite was steak and kidney pudding. Stews tasted great in the fresh air. Weekends, the menfolk visited their families. The grownups went to the village pub while we children played outside. My uncle bought me pop with a glass marble in the bottleneck and if I got it out he gave me a shilling. For a treat we went to Tankerton for a Knickerbocker Glory.

Shirley Thorn

Half-a-leg

One morning a young lad came out with one leg short on his jeans because his sister had torn hers and they'd taken the leg off his to patch hers.

Q. Moody

London Kids

When enough of the hops were pressed into pokes they were loaded onto a lorry and taken up to a London brewery. I went with them once. While waiting for the lorry to be unloaded I sat in the cab. There were these London kids climbing all over the lorry and I was too scared to move as I was only about twelve.

E. Grimwood

Nasty Surprise

Aunt Rose took us hopping at weekends. We picked all Saturday then explored and scrumped with the boys all Sunday. One day we found a field, all burnt. We went round in our wellies, jumping in the ash. Cousin Bob jumped with both feet, hot ash went over his boot tops and burnt him. We had to carry him back to the huts.

Anne Durrant

Little Lamb

I'm the eldest of ten children, all born in twelve years. Mum called it our holiday but it was no holiday for us. Life was hard. Dad, a docker, rarely worked a full week. Everything was rationed. We constantly queued for everything. On our farm, Cliver, where incidentally my mum's ashes were scattered, the shop was a big wooden shack. On the last day everyone packed their cooking pots into the old hopping box my dad made. They put me in the queue as we heard the butcher was giving extra rations of meat. I'd been told to get a leg of lamb and 2lb of sausages to take home. But with all the packing done everyone went to the pub to wait for the little bumper train taking us to the main London train. Three children were hid under the seat from the ticket collector. Well everyone forgot about me being in the queue and dad having a couple of pints, poor mum wasn't used to even one drink and my dear nan had been bought a few, so my dad had to try and get them, the luggage and all the family down to the station. He eventually got nan and mum a seat, then as the train started to move they suddenly remembered me still queuing for meat. Dad came running up the hill to meet me running down holding on for dear life to this leg of lamb and sausages and they had to pull me up into the carriage as the train moved out of the station.

June Pearce

CHAPTER 6
Scrimping and Scrumping

Mrs G. Smith and her mother at the hop bin.

Apple Pie Bed

One year, me being the smallest, my brothers made me climb and do the scrumping. The farmer, Mr May, came round searching with the police so Mum put the fruit in bed and got in on top. My brothers said she was ill otherwise he'd have ordered us off the farm. Mum punished us by making us go out with the Salvation Army. One night my sister and I decided we couldn't wait for our mum and went off to the farm about midnight. We were terrified passing through the cemetery to get to May's Farm. I think we woke the whole Common up with our shrieks. Outside Mum's hut was a tree and we were told a woman hung herself on it. We swore we saw her. Mum made friends with the family in the house at the bottom of the lane and one night when two ducks went missing my sister and I had to ask her to cook them so we didn't have the smell hanging round the hut. Well you can imagine what Mr May would have done!

Ethel Chandler

Watch Out!

The locals or home-dwellers didn't like the annual invasion of pickers. The shops draped netting over their counters to stop pilfering and the pubs charged a refundable shilling deposit on a glass. Mind you, your glass had to be well guarded: any left unattended were soon whipped away by vigilant kids; a shilling was a lot of money in those days.

Joan Jeffery

John and Joan give a helping hand.

Hop Cure

On Saturdays I used to go round to the kilns nearby, where the hops were heat-dried with charcoal and sulphur. When they were done they were put into very large pokes, which are sacks about six feet long and a yard round, then they were pressed tight. After a few days a small, square sample was taken out. My father also took a sample home in a linen bag, then if any of us had ear-ache they'd warm the hops and hold them beside your ear. It was a very good remedy.

E. Grimwood

Maud's son, Leslie, evidently enjoyed swimming!

Cobnuts Galore

After dinner and chores us kids had a great time. We scrumped apples, making the boys climb over the fences before the farmer came. As it was a cobnut farm there were lots of pigs in different paddocks. I went off on my own with a small haversack, in among the pigs and piglets getting some of the cobnuts for Christmas. You had to be careful not to be caught, else it was 'Off the field, pack up and get off the farm.' Dad used to take home a bagful each weekend and they were put into a suitcase and left to dry in our loft.

Enid Styles

Apple Scrambles

On Wednesday evening all the children wanted an apple picking scramble. The farmer's son, an active man, carried a few baskets of apples. On 'Go!' we had to chase him down the meadow picking up the apples as he threw them down. It was all good fun and the farmer made sure all the children got some. At milking time we children stood at the open barn door, watching the cowman milking. His nimble fingers drew off milk into a pail between his legs. One day we saw some of the older boys the other side of the cow. Unbeknown to the cowman they poked the poor cow with a long branch through a gap in the barn wall. The cow reared up and over went the cowman and all his milk. Next moment he was up on his feet and chasing the lads. On Saturday Uncle pushed Jimmy and me in the cart to Marden to a cattle market. It was exciting with the crowd of locals and Londoners.

Bob Richards

Vegetable Stew

I must admit to taking a large bowl into the farmers' fields at night and filling it up with potatoes, swedes, carrots and turnips and taking it back to our mums to cook a lovely stew on the fire the next day. We were only one of many families that did it; those poor farmers must have lost pounds, but thank God we were never caught.

R. Vaughan

Annie and Sid with their family, June, Ann, John and Pat (later Mrs Facer), at Ploggs Farm, Paddock Wood.

'Bring Out Your Dead!'

We had a nurse came round once a week calling 'Bring out your dead!' and there was a Sunday service on The Green for the children. There was singing round the fire at night and we cooked potatoes in the fires. We played in the woods, collected nuts and went scrumping apples. The bigger boys built swings from the trees. At the weekend most of the dads arrived and it was off to the local for some of the parents. I remember one year the farmer gave us two haystacks to play on. We had great fun for a couple of days, jumping and sliding on them but by then they were flattened like a straw carpet.

C. Mortimore

Wild Mushrooms

We did what most pickers did over the years; raided the orchards for fruit and watched where the chickens went about the fields to lay their eggs in the hedges. If we found some we'd have them away quick. As I got older I got the job of going back to the cookhouse at lunchtimes to make cocoa and sandwiches. Sometimes it

There was room for the family pets to enjoy themselves, too.

Harry Thirkell and the family dog.

took me a couple of hours because I'd go fishing in the ponds near the cookhouse, then I'd make all sorts of excuses about how hard it had been to get the fire to pick up, or something. When the family got the sandwiches there was nowhere for them to wash their hands and the sulphur dusting and chemicals made them black and tasted terrible. I learned a lot from country people: how to snare, clean and skin rabbits, which helped out with the meat ration. We'd collect wild mushrooms and cooking apples. My aunt made the best apple fritters I've ever eaten, now or then. We used the same huts right up to 1954. We went down to the farm last year for a

visit and heard that the hop huts now have a preservation order on them.

Graham Turner

Perfect Red Apple

One man, Smiler, with a nice smile, said to my mum 'Let's go scrumping.' They went to this farm and she went up the ladder and Smiler said 'Come on, Flo, there's the farmer,' and he cleared off. But Flo saw an absolutely perfect red apple and she couldn't leave it, so she got caught. The farmer said 'What are you doing in my orchard?' and he made her tip all her apples out of the bag and leave them behind, so she came home crying.

Edie Mortimer

Paid Off

If you got caught scrumping you'd be paid off with the paltry sum you were due for work done, but had to go home and miss the rest of the season. Once when we went conkering a load of cows came across the field towards us. We thought they were bulls and ran for our lives. We'd dropped the conkers and my jersey with them so the next day I was sent back to collect it before we went off to do the picking. It was in a terrible state after being trampled on by the cows but mum washed it and I still had to wear it because it was my only one.

C. H. Whittle

Poor but Honest

Our local pub was The Star and at night we'd go up there to see mum. But we'd have to go past Cut Throat Alley where they said a postman had been murdered and we were that scared. My granddad Napp always walked backwards and he even pulled the pram with the enamel teapot and sandwiches in it behind him. One year my gran had the money from all the families' Christmas Club in her handbag and she put it on the pram for granddad to look after at the hopfields, but it fell off on the way and no-one noticed. There was panic later when she found out and Gran had to report it to Mr Mainwaring,

but someone had handed it in. He asked her what was in the bag and when she described it correctly he handed it to her. She was so pleased she bought the finder a drink. I liked to catch lizards and when their tails broke off I used to put them in the po under the bed. Mum would give me a backhander when she found them. The walls were so thin you could hear everything, even the po scraping on the floor at night.

Edward (Ted) Newton

Ferret

Jim Grey always took a ferret with him and kept it in a hutch outside his hut so he could get a rabbit for his Sunday dinners.

Stan Dalton

Chicken and the Egg

My brother and I used to go out about four in the morning to pick horse mushrooms. On the way back we'd cut through an orchard where the chickens were about. If we saw any eggs we would take them and if the chicken made a noise we'd take her as well, wring her neck and take her back to Mum.

Rosina Amis (née Rose Elliot)

Family Job

I can remember all the family going out to the orchards very early on some

A family of pickers.

mornings. We'd pick apples, pears, plums and whatever we could find. Not just us kids but the mums and dads went as well. Kent cobnuts tasted delicious. The fruit we picked was all hidden in your tin hut in case the farmer came on a search, which he often did.

W. Leman

Alerted!

At nights we sat round the fire chatting or walked to the 'West End' pub and some went scrumping apples. The security man came round one night and we gave him some spotted dick and kept him chatting so the scrumpers could get into their huts.

My uncle sometimes took apples home, made them into toffee apples and brought them back to sell. Once when he got off the tram in Old Kent Road, Dad's attaché case flew open and apples rolled along the road. Dad was so embarrassed he just left them. Mum was one of the first to have a Tilly lamp. We could see paraffin lamps shining through the gaps between the huts and they'd shout 'Turn up your lamp, Dolly!' so they could see better. Dad came down for weekends. The children went to the village school to see films. I don't think the shopkeepers looked forward to hopping time – they always put chicken wire in front of the counters to keep us off.

Joan Lewer

Patricia Facer's family in September 1951, including her mum (then aged thirty-nine), Aunt Mae, Alex, Ray, Philip, John, Louis and Towzah.

Saturday Special

Dad came down every Saturday afternoon, bringing the meat for Sunday dinner. It was usually a hock of bacon. We worked in the fields Saturday mornings but in the afternoon went to town to buy provisions then go to the pub. Once, Dad had a drop too much and injured his ankle falling in a ditch. He ended up staying there all night. Mum used to carry the vegetables home in her apron and once she dropped the lot. The boys had to go out next morning, find where she'd dropped them and pick up whatever they could find or we'd have been without dinner.

There was no Sunday opening for shops in those days.

Joan Jeffery

Depths of Hell

I used to be in the Cubs and wore a green jumper, but we were so poor that it was my only jumper so I had to wear it all the time. One time I went to the toilet – just a square tin hut with a wooden seat over a very deep hole. Because I wore braces, I had to take my jumper off and put it on the seat. While I was struggling with the braces I

The Jillings family take time out for a welcome cuppa.

knocked the jumper off into the pit, down into the depths of hell. We couldn't afford to lose it so my mother came along with a big pole and fished it out. She soaked in it buckets of water to get it clean again so I could go on wearing it.

Tom Gower

Calamine Lotion

I can't remember there being any serious accidents. We used calamine lotion for nettle rashes and such like. We couldn't get rid of the stains from the hops except by hard rubbing with dirt, grass or rainwater. The hardest part was walking in the rain and mud to the hopfields. If it was wet we wore any old

Hop-picking is fun for these youngsters!

Jean Pilbeam and her mum chat over a bine.

macs and hats while we were picking. Sometimes we had to pack up early if it rained all day.

Enid Styles

Cooking for Twenty-five

On one Sunday in the season the whole family would come down in an East Kent bus they'd hired for the day. I don't know how my gran managed it but the tablecloths were spread out on the grass, Granddad would bring down a salt leg of pork which he'd cooked before leaving home and Gran cooked mountains of potatoes and carrots for about twenty-five of us. She'd make a massive basin of custard and we'd have that with tinned fruit. She was a marvel, seeing as she weighed twenty

stones, but she enjoyed every minute of it and so did we. On the last Sunday the men used to put all the ladies into the bins as a big joke.

N. Burney

Blackout Bedding

Mum purchased some blackout material, which didn't need coupons, and spent day after day bleaching out the black, then hand sewing it up to make a mattress cover. When we went hopping we filled it with straw. Dad made our hopping box with removable handles and pram wheels so we could use it as a table in the hop hut, but during the winter it was used to collect coke from Vauxhall or Nine Elms gasworks and firewood

Jim Ash on Betts Farm, 1933.

kids its vicious teeth. 'They can bite your finger off,' he'd say.

Bob Orris

Snares

Traders came down from London and set their stalls up on the village green. They sold everything. Mostly the locals bought as we'd seen everything before. The gypsies were just inside the gate of our farm. They lived in painted wagons and sold pegs. One old lady smoked a clay pipe. There were six of us so we had plenty of boots and clothes between us. We just wore them until there was nothing left to mend. We put snares down and tried to trap rabbits but we never caught anything. I doubt if we could have eaten it if we'd caught it anyway. We never took anything other than apples.

Jean Pilbeam

Kentish Cobnuts

It was a magic time for me as a child, what with pinching the farmer's fruit and going nutting – that's not violent, that's gathering nuts from the hedgerows. When you think how unsanitary the conditions were that we lived in, there was virtually no illness and we thrived on it.

John Meinke

from bombed out houses. There was an impressive windmill at Cranbrook: the first I'd seen outside a book. We found mushrooms and searched for wood pigeons' eggs (smaller than hens' eggs) and had a good fry-up. The farmers said pigeons were classed as pests but we only took one egg from a clutch of two. We called the general store The Oil Shop. It sold paraffin, candles, snares, nets for rabbiting and fishing line which we greased with candle wax to make it float. Our floats were birds' quill feathers pierced through corks. One year dad caught a pike. He brought it back to the hut then gutted, cleaned and cooked it on an open fire. He showed us

Animal Sweets

In our field a man came round every day with a box of sweets on his shoulder. He had sugar rabbits, cats, dogs and monkeys on sticks for a halfpenny each. I used to ask 'Can I have a red cat today, please?'

Edie Mortimer

Scary

The older children and some of the grown-ups went scrumping in the dark to get fruit. The apples were massive. We'd pick blackberries as well and my aunt made them into a mouth-watering suet pudding. Once, I was up an apple tree throwing apples down to uncle when a lorry drove up and trained its lights on the orchard. We had to squash behind the tree trunks to hide, then stagger back to the hut with heavy bags of apples.

Shirley Thorn

John Thirkell rides round a finished field.

CHAPTER 7

'Sing, *Let the People Sing*'

A group of hopping ladies.

'When we go down hopping,
Hopping down in Kent,
See old Mother Riley
Putting up her tent, with a
Tee I oh, Tee I oh, Tee I ee I oh.

Monday is the washing day,
Don't we do it clean?
Boil it in the hopping pot
And lay it on the Green, with a ….

Every Saturday morning
Farmer, he comes round
With a bag of money
And puts it on the ground, with a….

Do you want a sub?
Yes sir, if you please,
To buy a hock of bacon
And a pound of mouldy cheese, with a….

Joan James (This song has many variations.
HH)

Hop, hop, hopping in the old Kent fields,
Pick, picking all day.
We rise with the sun
Work 'til the day is done,
Hopping in the old Kent fields.

Hop, hop, hopping in the old Kent fields,
Not far from London town.
We camp in open air
With huts and wagons there,
Hopping in the old Kent fields.

Hop, hop, hopping in the old Kent fields,
Among the leaves so green,
A bine beneath the arm
Gives just the girl her charm,
Hopping in the old Kent fields.

Alice Heskett

Old Songs

I loved the afternoons best. After a packed lunch and tea made on the meths stove the families would set to work picking again and before long someone would start to sing, followed by another and another, until the whole field of perhaps a hundred people would be singing 'Nellie Dean', 'The White Cliffs of Dover' or 'Sunny Side of the Street'.

Rita Game

Weekends

Someone played a piano accordion and mouth organ outside the huts. Dancing on the grass wasn't very elegant but it made for a lot of marvellous fun and laughter. We played darts on a dartboard hung on the side of a hut and played cards and dominoes. Scrumping was a must for the children.

Jim Wood

Ditched

We picked hops from eight in the morning till six at night. After dinner we'd all sit around the fire for a sing-song but at weekends they'd walk about a mile and a half to the Peacock, which was the nearest pub. People in the families who still had to go to work came down at the weekends by lorry for a five-shilling return fare. It was pitch dark walking to the pub and back. Some used to fall in the ditches and couldn't get out for laughing. Most of us had hand torches. One friend was visiting one weekend and he got a bit the worse for drink. He was missing from his bed and they didn't find him until next morning, asleep in the grass outside still with his torch alight.

V. Hammond

Real luxury!

It was a trudge back to the huts after we'd finished picking and our priorities were food and wash in that order. When we'd cleared up we'd sit round the fires or on the common singing, telling stories and having a good chat. If there was somebody could play a mouth organ or piano accordion that was a bonus. Gramophones were expensive then but when they were more affordable in later years we took a portable wind-up gramophone. Occasionally the Salvation Army visited us, usually on Sundays, and they showed us religious films from the back of a lorry. We were too far from the town to go to a pub in the week but we went at the weekends when visitors came down, mainly fathers. In some ways this was just as well as some people would have spent all they earned each day.

Joan Jeffery

Old Songs

Mum died three years ago. She was ninety-six and the doctors were

amazed at how well she kept. She used to sing 'A Rose in a Garden of Weeds', 'Row, row, row the boat,' 'Ramona,' 'All the Little Pigs Run with their Bums All Bare' and 'The Spaniard who Blighted my Life'. My sisters are in their eighties now.

Edie Mortimer

Fire and Lamplight

We picked at Tewdely Church Farm near Tonbridge, owned by Mr Young. We also went to Homestall Farm which was owned by a Mr Bones and Mr Thinn. I'm eighty-three now but I was eight when we went. The pub nearby was the Carpenter's Arms. There were roses all round the door. I was with my mother but my grandmother took us. She had a family of nine. We went basket picking, not bin picking; it was more fun and better paid. Night came early in autumn, and after the day's work we'd sit round the stone-built cookhouse fires, talking and telling stories. There was no electricity; just the flickering fires and soft light from the hurricane lamps to see by.

Charles Whittle

Favourite Pub

When we were children we all got together as families and went to Homestall farm, hop picking. We were a very large family so this was our holiday. Our last time was in 1956. In the evenings we'd all walk to the Duke of Kent pub which was a meeting place for

Rosina Amis's family and friends enjoy an evening at the pub.

all the hop picking families. Someone would play the piano and we'd sing all our favourite songs, like 'Star of the Evening'.

Joan Kendrick

Spontaneous Band

In the evenings there was a gang of us round the fire, talking, singing and having games. The men would play the spoons, the mouth organ, accordion, or tin drums – anything that would make a tune. Just good, clean fun. Those were lovely days; everyone so friendly. At the end of the season we all joined in a big party round a bonfire. Sometimes the farmer and all the workmen joined in, not that every place was the same.

Elizabeth Webb

Worker's Playtime

The pole pullers sang all the latest pop songs. We'd sing round the campfires of an evening and tell eerie stories and jokes. Our binmen and pole pullers came from Notting Hill and were brothers. Each night crowds of kids would sit around the fire on our stools and boxes singing along with them with their guitars. There was 'I Believe', which was Mario Lanza's, and Tom Jones' 'Jezebel' and the latest Johnnie Ray hits. Frankie Lane's 'Wild Geese in the Sky', 'The Man from Laramie', and all the hits they sang. In the dark evenings we all had torches in case we had to troop down the woods to the lavs; it was very scary if you went on your own.

Enid Styles

Three members of Florence Wright's family, Billy, Roger and George.

Flickering Firelight

It was great fun at night, sitting round the faggot fire. The faggots were delivered every couple of days by tractor by farm hands. We'd have a mug of cocoa and tell stories until about ten o'clock when the fires were almost burnt out. The huts we lived in were quite comfortable. We sometimes stayed round the hut area in the evenings and played cards, or went for a stroll in the High Street for a drink. There were three pubs in Goudhurst, the Vine and the Five Bells were two of them. We mostly went there on Saturday nights and all the Londoners got on well with the country folk. It always turned out to be an enjoyable evening although it was a long trek back across the fields. We'd all be chatting and laughing, but we pickers all kept together going home.

Doris Turner

Crisp Wages

We were allowed into Faversham on a Saturday afternoon and I saw my first picture there : *Old Mother Riley*. Once, when I was fourteen, I was playing the piano for the charabanc trippers in the Duke of Kent pub with a backing group I think was The Bachelors. Mum came hotfoot for us. I was upset as we were going to be paid in crisps and pop.

Doreen Dillon

Singing Along

The happiest times were at weekends when all the families came down. We'd all go off up to the village green to the pub for the evening. When the publican turned us out we'd all walk back down the lanes, singing all the way

Hoppers were often mistrusted, and had to pay a deposit on their beer glasses.

Rosina Amis's family at the local pub.

to the hop farm. There wasn't any fear of getting knocked over in those days.

T.W. Ovenell

Gramophone Man

My mum was always called Aunt Madge, even by Mr Chambers, our farmer. Our local pub was the Chequers (at Laddingford) and hoppers had their own room. There was another room for home-dwellers. The Salvation Army showed magic lantern shows in a little hall and there was a nurse had a little hut where she looked after the hop-pickers if they had any injuries or cuts. One night when we went to the Chequers there was this man walking along with his gramophone still playing. He mustn't have seen the dustbin because he fell over it and the gramophone scattered all over the place. He really turned the air blue. There was one hopper from Hackney, Ginger Yellup, who used to play the piano really well down at the pub. Hoppers had to pay sixpence extra on their beer for the glass. If you took the glass back you got your sixpence back. On Sunday mornings we used to go round outside the pub and collect all the glasses we could find and get all the sixpenny deposits back from the landlord. There were so many hoppers they put planks across wooden barrels outside the pub for seats. At Buster Manor in Yalding there was the gipsies' inn. The Travellers

Joan Thirkell with her nephew at the hop huts.

camp fire outside the huts, singing or playing.

W. Leman

Down to a Fine Art

There was a severe lack of privacy, but in most cases the hoppers seemed to have the crude cooking facilities down to a fine art. At first I went just for a lark. I remember once there were a lot of wasps came from nowhere and went for our jam sandwiches. Whatever food we got we were grateful for; it was better than army food and the smells were gorgeous. In the evening I went with the hop-pickers to the pub and it was all jollification and singing. I enjoyed my shandy. Some of the songs were 'There's a quaint caravan and the lady they called The Gipsy…' and 'It makes no sense, sitting on the fence, all by yourself in the moonlight'. The pickers were honest as the day was long and good company. They must have handed the whole operation down from mother to daughter and generation to generation as everyone knew just where they were headed.

Jim Vansen

weren't allowed inside the pub though. They'd have to send someone in with a big jug, then have their drinks outside. Opposite Chequers were allotments and we could buy our veg there. We sometimes went to Paddock Wood to a pub called the John Brunt. He won his VC at Anzio and there's a monument in the village to commemorate him.

Arthur Smith

To Each his Own

On Saturday nights all the men went to the pub for a drink and all the mums and kids used to sit around the

Hymns

We didn't pick on Saturdays or Sundays. The church held a meeting on the green for any denomination and all the children would go. The singing could be heard all over the fields.

A. Hamlin

Jean Pilbeam's father, Tom Bird, relaxes at the end of a long day in the fields, 1938.

Hitler Preacher

A religious group used to come to the common every Sunday. I remember my brother, who was only eighteen months old, was terrified of the preacher who had a beard and wore a hat similar to the one Archbishop Makarios used to wear. My brother used to cling to my mother's skirt and scream 'Daddy, Hitler's here!'

Patricia White

Looking After the Aunts

We went to Chambers Farm about one and a half miles from Horsemonden. There was a pub in the village called the Woolpack and my grandmother used to frequent there. She was Mary Stamp and her son is Terence Stamp the actor; he's my cousin. Grandmother went with a distant aunt of mine called Nina Ford and my mother sent my cousin and me off to collect both of them from the pub at night. We took a torch because there wasn't any street lighting or anything. It was really embarrassing for us because they were always half cut. We were only twelve at the time. They used to wear long skirts and those long-legged knickers and they'd stop by the side of the road, pull up one leg of their knickers and urinate, letting go like young horses. At the end of the six weeks or so of hopping they'd not much money left because they kept subbing for the booze.

Tom Gower

97

Irene Crimins (second from the right) at the machine shed, 1976.

Little Hoppers' Hospital

Just along the road from the village was the Little Hoppers' Hospital and once a week two lady helpers used to come and collect us children and take us there to see a magic lantern show. The same ladies took us to Sunday school and gave us religious stamps. If we had a full card at the end of hopping we were given a small gift. They came round the hop gardens with a tea trolley and sold cakes and things. A gipsy family in the field next to ours made toffee apples while we all stood there watching with our mouths watering.

Irene Crimins

Father Richard Wilson, Stepney

There was the Hoppers' Hospital at Five Oaks Green where we went if we'd hurt ourselves. A Stepney priest, Father Wilson, bought a pub and made it into a hospital because he was sorry for the hoppers who couldn't afford to go to the doctor if they got a nasty injury. But it was him who got the farmers to make our huts better. They gave us any old thing before then; even sheds the animals had been using for winter shelters. Funny, but the little hospital still looks like a pub.

Nell Hearson

Hoppers' Pudding

We went to Hawkhurst where my wife's family went for years. I remember a character there called Nell. She used to cook a huge apple and blackcurrant pudding every weekend and everyone had a share. We had visitors down at the weekends and we all went to the local pub in the evening. When we got back to the huts we sat round the fire and sang songs. One year big Nell brought a piano down with her on a lorry. When hopping was finished we had a real ding-dong of a party and afterwards we burnt the 'joanna' on the fire because Nell didn't want to take it home.

F. Bowstead

Home-dwellers' Evenings

At the weekends there would be a big bonfire with a lot of songs and I suppose turns, but I can't remember our lot mixing with the Londoners (we lived near Chatham). The local pub did good trade but the pickers weren't allowed inside and were served from a window. They had to pay a deposit on their glass.

Mrs M. Murray

Car-eering Conveniences

People came down for weekends in their cars. The atmosphere was so lovely, everyone happy they were earning a little extra. Anyone owning a car was posh to us. Well this

At Seal, near Sevenoaks, in 1949.

particular Saturday afternoon the men had had a few drinks, the women were doing washing in tin tubs and the children were in a big marquee and you took your own stool to sit on and watch a film on a large screen. The children loved it, so did my nan. Anyway we were round the open fire singing songs then suddenly we saw this car rolling down the hill which our huts were at the top of. Its brake had been left off and as it got faster we could see it heading towards the row of tin huts which were our toilets. Well next it crashed right through them. When it stopped there was this man, sitting on the toilet, trousers down but, thank God, not hurt, just shook up. Well, you can imagine the load of

The Hams family outside their hut, beside a hop field.

people watching what happened as the camp held about two hundred tin huts. I'll never forget that Saturday.

June Pearce

Aunt Bob

Aunt Bob was always the joker in the pack. She had everyone in fits of laughter with her impersonations of Gracie Fields, marching up and down the drifts in wellies singing 'Sing as we go, and let the world go by', or trying to walk on her knees with Rose Lynch holding her hand as if she was a child, singing 'I'm a sister in the Salvation Army'. She'd get the whole set singing

popular songs of the day like 'My old man's a dustman' with a few rude words thrown in. One night she woke up and saw the hop hut door swinging open. She looked out into the hopfield and there was this shadowy white figure in the night. She was petrified and called out to her daughter-in-law Rose. As she was calling the ghostly figure walked towards her and Aunt Bob was about to scream when she saw it was Rose, coming back from the field after answering a call of nature.

Bob Orris

Grave Apparition

We children watched all the old films – mainly *Old Mother Riley*. After the film we'd walk home together. It was over a mile and we had to pass alongside a big graveyard. Once we saw a white apparition swirling through the gravestones. Although we were all brave when we were together, when some of the group started running the rest of us lost our courage and followed after them. Now, several years later, I think it may have been just a wisp of smoke… or was it?

Michael J. Stark

He Enjoyed the Beer

The fathers were usually at work in London during the week and they came down at the weekends. My dad took his annual leave during the hopping time and as a treat he liked to take me into Maidstone for the day.

Michael Stark with uncle Maurice Dunster at Whitbread Farm in 1941.

Fred Presley, Percy Piper, Bert Crook (who had a newspaper stand in the Strand) and Uncle John Neville enjoy a drink.

We'd sit by the Medway and eat our sandwiches by Fremlin's Brewery. Dad didn't like picking much and his contribution was to drink the stuff the hops produced. After work the only real form of amusement was to go to the Chequers which was our local in the village and we'd have a good old sing-song. The locals probably breathed a sigh of relief when we all went home – although they were glad enough to take our money. There were fights and drunks and it wasn't only the men!

John Meinke

Sugar Animals

All the men came down at the weekend. One man gave my dad a go on his motorbike. My mum was a bit of a goer, always singing and joking. Once, she was telling someone else's husband a joke and his wife got a bit jealous and came out with this bag of flour and threw it all over him. After all the little ones had been put to bed our parents went off to The Forge. We were given a lemonade and big oval biscuits called Brighton Biscuits. They cost a halfpenny each. Every now and again on a Saturday the men had a day's outing in a charabanc from Callender's

Cable Works. All the men had already been in the pub for a couple of drinks so were well away. We children stood against the fence and when they came out they waved and called out to us and threw us halfpennies and pennies.

Edie Mortimer

Accordion

In the evenings we sat round the fire singing the old songs while someone played the accordion. We children weren't allowed to wander off. We lived in Picardy Street, Belvedere and picked at Merriworth.

Nell (Alice) Smith

Len and John Thirkell on an August Bank Holiday. The camp toilets are in the background.

'Twacker' Barlow

One year we met a young porter at London Bridge station called 'Twacker' Barlow. He was on probation for some minor offence at the time and had to work on the railway. He found us an empty carriage and at the last minute decided to travel down with us and stay. My mum had to send a telegram to his mum at the first opportunity so she'd know where he was. One day we decided to go fishing over at Bedgebury School Lake. The girl students were still on holiday so Twacker decided he'd go for a swim. He stripped down to his underpants then climbed a tree with an overhanging branch that was about four feet over the water. He gave a blood curdling Tarzan yell and dived into about two feet of water and two feet of mud. He came up looking like Al Jolson! Luckily he wasn't hurt but I think his ego was slightly dented.

Bob Orris

CHAPTER 8

War in the Hopfields

Shirley Whiting's dad, with their old dog, Judy.

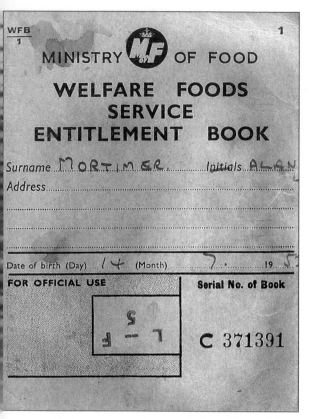

| WFB 1 | MINISTRY **MF** OF FOOD | 1 |

**WELFARE FOODS
SERVICE
ENTITLEMENT BOOK**

Surname MORTIMER,   Initials ALAN,

Address ....

....

....

Date of birth (Day) 14 (Month) 7. 19 5

FOR OFFICIAL USE   Serial No. of Book

C 371391

Rationing was still a part of life well beyond the end of the war.

The Day War Broke Out

I was fourteen when war broke out and we were held up all night on a hop pickers' train going down to Cranbrook. No-one knew what was happening and it was only when we got to the farm we found war had been declared. The farmer spent the day making us practice with our new gasmasks which had been collected from Dockhead.

Bob Heather

War End

As the war was finishing my aunts, uncles, cousins and my own father gathered together, celebrating the fact that they'd all returned safely from Europe, India and the sea.

Jim Wood

Country Residence

I went to Charlton Manor School before the war but they closed it as most kids were evacuated. We were the only family left in our street during the Blitz who still went hopping. The teacher came round and we went to a house down the road and sat at a table trying to learn. Then the air raid siren went and we went back home to our dugout. Our neighbour across the road said she went to her country residence on Friday nights when bombing was bad, then we found they went down to their own little patch at Chislehurst Caves. Enemy planes came in low over the hopfields but most were heading straight for London so there was no need for shelters and we felt quite safe. I can't remember there being air raids at our farm, Highward's Swigs Hole, Horsemonden.

Jean Pilbeam

Spy

When the London bombing was intense more people than normal came down hopping. I made friends with one woman who didn't seem to

Jim Wood (small boy) and his sister (in the gym-slip) with their parents at the hop bin.

have any belongings. I went round different families, collecting a kettle, blanket, comb and some bits including my nan's hairpins. After three days I went round to see her and was told the police had taken her away: she was a German spy and had been caught signalling to a plane that had circled for a couple of nights. I never heard the last from my nan about the hairpins.

Joyce Ashby

Great Times

In the war it was the only place where you could say you were going to get a good night's sleep. My dad used to go outside on Sundays and shout 'Wakey-wakey, rise and shine!' Some of the pickers came from Tilbury and Grays in Essex. Our Aunt Jessie played the piano in the village pub every weekend.

Mrs W. Fagan

Lord Haw-Haw

Sometimes the hops were small and when the measurer came there wasn't much in the basket and your pay was small although you'd done all the work, so this sometimes led to strikes. All cooking was done outside. Tea was made in a big black kettle with a twig in it to keep the taste of smoked wood. The last year before we got called up for the war there were only about six of us, no women at all. We saw aeroplanes fighting above us every day with German planes crashing everywhere. It was frightening. We went to the pub a

105

The hops were especially good in 1953, and Enid Styles' mum and dad and Enid Woolterton show their appreciation.

couple of times a week and heard Lord Haw-Haw on their wireless, saying 'We know there's a lot of people in the hopfields, so the German planes will get you.' This didn't frighten us, but one night it came over the wireless that the Surrey Docks and the East Coast were bombed very badly and we worried about our families so returned home. We both got married, then got called up for four years. My wife came from Bermondsey. Her father and brothers worked in the Tooley Street Docks.

Mr C.H. Wiggins

Break from the Bombing

We went to Wateringbury during the war, in 1942. My mother and sister-in-law went down for a break from the bombing. And this time we had nothing with us so the beds were very bare, just the straw, and with rationing we'd hardly any food, though there was a pub, the Iron Duke with Two Heads. We weren't allowed any lights as we were very near a searchlight unit and an aerodrome but we'd sit outside the huts in the dark and tell stories. I was only seventeen and met a good soldier and would you believe it, he came from Bermondsey, not far from me. He kept us alive with his jokes. My soldier used to come to the bin and help me pick so I could go out with him and he took me on to Tunbridge Wells to the pictures. What a treat!

A. Hamlin

Ethel Chandler's family at Seal, Sevenoaks, in 1949.

Salutary Lesson

About 1940, I went picking somewhere near Lewes. It wasn't so friendly there as other places we'd picked. When the siren went we naturally dived down by the bin when we saw the bombers coming. The villagers laughed and took the mickey. Then one night there was a really bad air raid and one big bomb dropped in the village and another two outside. Nobody was hurt. The next time we were picking it was they did the running, not us. That was the lesson they'd learned. That was my last visit to a hop garden. I'm eighty-four now.

Elizabeth Webb

Shrapnel on Huts

I remember hopping during the war and lying in our tin huts listening to the German bombers going over to the Channel. Lots of shrapnel fell on our hut roof and in the morning us kids would go out and collect it. Sometimes it was still warm and we'd see who could find the biggest piece. It was exciting to us, but not to our poor mums who must have thought it was hell. But even Hitler didn't stop all the lovely singing and laughter in the hopfields.

R. Vaughan

Soldiers Three

My brother Jack joined up before conscription came in. He came

Shaving in the morning wasn't easy!

victory roll over the gardens all the hop pickers would cheer because we knew a German aeroplane had been shot down by the pilot doing the victory roll.

Patricia White

German Tallymen

In the earlier years German prisoners of war supplied the wood for fires and weighed the hops: I believe the going rate was 1s a bushel. One particular prisoner, Hans, was very friendly and jovial and always talking to the pickers about the day when he would be repatriated. We spent Friday nights in Paddock Wood at the 'twopenny rush'.

Michael Stark

down to help with the hop-picking when on leave and he was so bronzed the others thought he was a German POW. My youngest brother, James, was in the Guards in Cyprus and came picking with us during 1945 and 1946 after leaving the army. Charles joined in 1935 and was a POW in Japan. He didn't come back until 1947.

Charles Whittle and Joan Kendrick

Victory Roll

Paddock Wood was quite near Biggin Hill and we'd watch the Hurricanes and Spitfires going in and out of the airfield. If a plane came back and did a

Steady Buffs

I was with the Buffs and stationed all over Kent including Canterbury. I played football for the Army and didn't have to do square-bashing as I went cross-country running so was often in the hopfields. London seemed to be ablaze when I went on leave. You didn't know if you were going to be alive tomorrow. I had a great friend, Reg McLaren from Dockhead, and when I came out of the army Reg and his family took me down to Wateringbury every summer in his 3cwt ex-army wagon. I must say most people would have flinched at spending weeks in the harsh living conditions: compared with the army they were really going to rough it!

Jim Vansen

It's dinner time for Mum, Rene, Alf, Daisy, Jean, Teddy, Brian and Derek.

That Sunday, 1939

We came home sadly but the money we earned set us up with clothes and what we needed for the winter, but it ended for us in 1939 when Mr Chamberlain declared war on Germany. We sat outside the café – it's still there by the Medway – and heard him on the radio.

John Meinke

Bombs Away

My husband's family used to go to Staplehurst in the 1940s to get away from the bombing for two or three weeks. They had a lovely time at Rose Villa on Day's Farm. They shared with three other families.

V. Hammond

Hasty Retreat

My mother absolutely hated the toilets and one day when she was in there the Germans came over and machine-gunned the common (not so funny at the time). My dad shouted at her to stay where she was as it was safer, but no way would she! She made a dash back to the hut rather than get bombed in a toilet.

Irene Crimins

Bombed Bunnies

Our parents first took us in 1935 and we continued to go hop picking until 1959 when I was married with three children. Except for one year we always went to Angley Farm in Cranbrook. I remember, about 1943, it was a very dull, cloudy Saturday afternoon and an aeroplane had been flying around which we all thought was a German bomber because no warning had gone. Then suddenly there was a burst of machine-gun fire and the whistle of a bomb falling which sounded very close. Dad shouted 'Get down!' and everyone lay down on the ground but it actually fell in Angley Woods, half a mile away. The following day a couple of fellows came onto the farm trying to sell us rabbits that had obviously been killed by the bomb, but they didn't make a sale. One woman said 'It's not your bomb-killed rabbits we want, it's some shell-shocked eggs' which were in short supply at the time. But we all had to use powdered eggs.

Ron Bird

Adventure

There was a poster: 'Is your journey really necessary?' My sister did her bit in the Women's Land Army and there was a shortage of labour on the farms so there was a Government scheme where workers in various industries could spend their holidays working on farms. I applied to work in Kent. It was something of an adventure to me and a chance to 'do my bit' for the war effort, so I not only got a holiday at Linton, but got paid for it by the farmer as well.

John Meinke

Italian POWs

Our faggots were cut by Italian POWs. They wore grey work clothes with a large black circle on their backs and were an amiable bunch, waving and calling to us on their way back in the evenings. All the meadows around Cranbrook had poles with wire strung across to prevent enemy planes landing. Hops were either 'peanuts' (small), 'alright' (average) or 'hanging like pears' if big. If the crop was small there were many disputes over the rates of pay and sometimes there'd be a strike. Once my dad was elected spokesman. He was a quiet, mild-mannered man but quite articulate. The farmer accused him of being an agitator and he was barred from Forge Farm, but he managed to get into Smugley's then back to Forge after that. My ninth birthday was on 26 September 1948 and when we got home from hopping there were my call-up papers waiting. Somehow the War Department had my age down as nineteen. We had the police round, later, and it took a lot of convincing there was no fiddle going on.

Bob Orris

Deserters

We lived at 44 Willow Road, Dartford and always went to Denison's Farm at Shoreham. We were there during the war and I remember the police raiding the hopfield looking for deserters from the army. Seeing them coming, a couple of ladies put their husbands in the hop bins and covered them up with hops so the police wouldn't find them. One night bombs were dropped whilst we were still sleeping and the next day we were all put into the lorry straightaway and taken home.

Stan Dalton

Bullet Pudding

My mum took me hopping at Faversham when I was six months old and my gran, auntie and mum had been going way before that and through the war. It was a big adventure to us children; we didn't see the fear our parents shared. We had some lucky escapes in the war. We'd just left one hopfield as the siren sounded and were hurrying back to the huts when the place shook with this heavy bang. When the all-clear went we went to see what had happened and there was a huge crater just where we'd been picking hops. Another time when we were picking, the siren sounded and Nan told me to take the children and keep them in the hut. I was the eldest at nearly twelve. Anyway I got them all sitting on the bed and we heard this aeroplane. A couple of my cousins were crying so I lay them down and lay over them then there was machine gunning; a noise we'd only heard in the films. It was so loud and the bullets came through the back of the hut where we'd just been sitting. One of the bullets hit my brother's furry toy dog and another had gone in my baby cousin's big gasmask. When another cousin found a bullet in his bread pudding everyone wanted a piece. The plane came in so

Twins Dolly Thirkell and Diddy Ford with Bert and the farm's cart horse.

low it took the tops off the raspberry canes.

Joyce I. Ashby

Doodlebugs

During the war my brother and his friend Siddy used to stand on the top of the cookhouse and watch the doodlebugs go over. Only one time it came too close and blew them off the roof. One of the local girls, Pam Rich, married Gerhardt, a prisoner. He changed his name to Gerald and they had two daughters. He made them

111

lovely toys. Wilhelm was a seventeen-year-old prisoner. He was some mother's son, poor little thing.

Iris Hans

Toads in Trenches

I had an aunt and when they all went down to the pub of an evening after the picking she got so drunk they had to bring her home in the wheelbarrow. We were at Paddock Wood during the war and they'd dug all trenches round the fields for the hoppers to shelter in during air raids. We kids jumped in and ran all along them until we found the toads and the girls all screamed! The boys thought it was great and used to throw them at us. I remember a German plane being shot down in the field right next to us.

Kitty Finch

Ninety-Three

When war started most of my brothers joined up. We lost some in the trenches. I'm ninety-three and have seen all my brothers and sisters out, bless them.

M. Sanders

War News

We took any food down to the fields that was handy for our meals and made tea and sandwiches on the spot. The men weren't ashamed to cook when it was out in the open and Uncle Albo liked to cook us pigs' trotters. We'd walk home from the pub singing because the country lanes were so dark and a bit frightening, but it was a lovely sunny day when my mum heard war declared on the radio and she rushed out into the field and brought all the children in straight away.

Marian Richardson

Grandstand View

Autumn school terms were adapted to fit in with hopping, but during World War Two not many London hop pickers were available from the East End, so a party of us living in Tonbridge answered an SOS from a farm in nearby Leigh village for pickers. We had to catch the train from Tonbridge. We had a grandstand view of many dogfights over the fields and occasionally took refuge beneath a railway arch. One day a low, swooping plane with a large black cross on it sent us hurrying to the nearest ditch. It was being chased by our fighters and they forced it down intact near The Old Barn Teahouse at Hildenborough. The owner was a bit of a wag and he put up a notice saying 'Landing ground for light planes'. In earlier days people used to gather on the verges of Riverhill near Sevenoaks to watch the pickers' laden carts with their horses and skid pans going down this very steep hill on their way to the hop gardens.

Ruth Grey
(The teahouse is called 'Oceans of Cream' nowadays but it's still there. HH)

Richard Carpenter (marked with an X), owner of Reeves Farm, and a group of hoppers give tea to a Spitfire pilot (left), after he was shot down.

Friendly Airman

We were two years into the war and one September there was a dogfight overhead between the Germans and the Spitfires while we were picking. When one Spitfire was hit the pilot bailed out and landed in the next field. People made him tea. Some took photos. He said next time he went up he'd fly over the hopfield. Two days later a Spitfire flew over doing the Victory Roll for us. We were sure it was the same pilot. All through the war we watched the planes crash and the lads would be in a rush to strip pieces off them.

Graham Turner

Flying Fortresses

Our farm was Hall's Poultry Farm at Marden. I think Mr Hall employed about a hundred families. Everyone knew everyone. The day war was declared me and the other children had been sent to town to the baker's with meat and potatoes in a baking tray for him to cook for us. Going back to the huts people were standing outside a home-dweller's house and we heard the Prime Minister, Mr Chamberlain, say that Britain had declared war on Germany, but being children we didn't really understand what it meant. We had only been picking a short while when German planes started to raid London and our planes were firing at them. We called them dogfights. We didn't have any fear, just sat on the common, watching them. We'd count the Flying Fortresses going out and we'd count them coming back home and my old gran would say 'There's four missing.' We'd be picking and someone would say 'Be quiet!' and the whole hop garden went silent. Then we'd hear the engine of the plane as it flew over towards us and the whole field of hoppers erupted with cheers and we'd go mad. All the women had tears in their eyes.

Rosina Amis

Our Lean-To

We had to go into trenches in air raids on Paddock Wood. We were at Matfield. The trenches were dug around the hopfields for the pickers to take shelter in but after a while the pickers carried on picking during the raids because they didn't get paid for not picking. We saw planes high up in the sky dogfighting. Some German bombers came down low and dropped their bombs as they were being chased by British fighters. I remember our dad building a lean-to in the local woods for us kids to sleep in of a night. He said it was safer for us during the night raids. After the war we still went picking now and then, but we stopped when I got married and had children.

F. Bowstead

Doodlebug Watching

We were all very good friends and got on well together. The night before we came home there was a pack of German bombers came over as we were walking down the field and a lot of our boys came up from an airport at the Weald and made rings round the bombers. The children were cheering as they watched a plane come down in flames. It landed about two and a half miles away from us. Our lads were marvellous. They turned back to refuel. They brought down about seventy German planes. We watched the doodlebugs come over in the night with flames coming out of their tails and saw them go over London. We were told that a few miles away was the place they called the German Graveyard. They managed to bring them down as they came over the sea. It is something we'll never forget and never want to see again.

Q. Moody

Pay Day

We weren't paid for our picking until the end of the season. On the morning of departure each bin holder took their tally card up to be checked for the amount. The card showed the amount of bushels picked at each measuring. If you were destitute you could sub, but this wasn't encouraged. The rate per bushel varied from one season to another. I remember one season we were paid a shilling for five bushels. On one occasion the pickers were dissatisfied with the rate on offer and they all came

Edie Mortimer remembers this VE Day street party in Barnfield Road in 1945.

out on strike until a more favourable amount had been negotiated with the farmer.

Joan Jeffery

Hoppers' Strike

About the mid-1940s we went to Chambers' hop garden near Southfleet. I recall it was a poor year for hops and the farmer was only offering 7d a bushel. Soon after we started picking it was obvious the small hops meant it would be hard to earn any money and we sent a deputation of hop pickers up to Mr Chambers to ask for more per bushel, but he refused. There was a bad feeling when some of the pickers refused to join in the strike, threatening solidarity, but eventually we reached a compromise and we went back to work. I can't remember the price we settled, but I do remember my picture being in the Gravesend Reporter.

Mike Pullen

CHAPTER 9
Happiness is a Hop Garden

Ann, Mary, Tricia and Josephine Gower waiting for the lorry to take them home to Poplar, 1966.

Local Pickers

We lived at Hextable and in 1930 my cousin and I went picking with a local lady at Shoreham. We were only walking distance from the hopfields. We put our contribution into an upturned umbrella then added it to her container. We had sandwiches for lunch and the taste of hops on our hands transferred to the bread and we quickly learned to wrap paper round the edges to avoid this. Some jolly folk from London made this their annual holiday and lived on site in wooden huts. They were happy, industrious people, often singing as they worked. They probably found the extra cash a help. I look back with pleasure and am glad of the opportunity.

Evelyn Knock

Sad Goodbyes

When hopping was over each year it was sad to say goodbye, but we knew there'd be next year to look forward to. Everyone helped each other and the kids were happy to be out of London, just messing about in the fields. I'm glad I was lucky enough to have been through it and to have so many treasured memories. We still have a day at Horsemonden every September and a meal in a pub to revive old memories.

Jean Pilbeam

All Changed

My family and I loved every year of hop-picking. We still have photographs that bring back great memories and we still go to see the farm now and again, but it's not the same any more. It's a shame that many farmers had their livelihoods taken away and it's all due to pressures of industry.

Mr and Mrs Andrews

Their Only Holiday

Really, it was the only holiday available to the poor of the East End. Now and again I drive down to Yalding; it's changed very little over the years. The Anchor pub is still there by the Medway, over the footbridge. The boatyards and the fishermen are still around. It's like seeing ghosts of people long gone, of course with mechanization there are no hop pickers any more. When I look back it was a lot of hard, dirty work and it was poorly paid. But I wouldn't have missed out on it for all the Costa del Sols!

John Meinke

Happy Memories

Not only us but our whole street in Camberwell went – almost all our neighbours. We took everything we could carry by lorry to the farm for the whole of six weeks' summer holidays from school. I'm fifty-four now, but

Tired at the end of the day!

I've very fond memories of hop-picking as a child. I enjoyed every day of being there.

W.D. Leman

Real Fun

We never wanted to come home after nearly a month in the country. It was real fun seeing the same people every year. I wish I could still go down hopping. I'm afraid the machines put a stop to all that, sadly. I'm in my seventies, now, and I've got some very happy memories of it all.

N. Burney

Punished

When we were at the farm I missed mother's home cooking as my aunt cooked on the campfire. But then there was our journey home and school again. The next day, those of us starting late back at school had to go up before the headmaster. The boys not with their parents, he said, were playing truant in school hours and would be punished. So he gave us all the cane. I missed my teacher at Riley Street School. He was jovial and happy, George Harris, but was due for retirement. We boys called him 'Harry Boy' Harris. He was strict, but a fair man. He made us work hard at lessons but afterwards he'd chat and joke.

Bob Richards

Happy Hoppers

I still go to any days in Kent such as the Faversham Hop Festival in August. Only this week I was listening on the radio to a programme about

118

Doris (right) and her mother-in-law (left) at Goudhurst in 1948.

Annie and Pat in their Sunday best in 1953.

Whitbread's. They'd put in a play area for children and were going to call it 'Mad Hoppers' so I phoned them and said 'Why don't you call it "Happy Hoppers" because that's what we were.' And they said they thought it sounded a better name so that's what they're going to call it.

R. Vaughan

No More Singing in the Dark

One year it poured heaven's hard with rain and the farmer had to put clinker outside the huts because of the mud. And then the rest of the time it turned out so hot they had to put up the big hop sacks for awnings outside the doors of the huts. I always remember the lovely smells in the evening down there when the fires were lit and people started cooking their dinner over the open fires. I went for a drive down Lenen Mile Lane recently and knew I was almost there, but they're making it more like a race track every day and there's no more walking down the lanes singing in the dark.

T.W. Ovenell

The Westcott family at the hop bins.

Flo Smedley with her daughters Pat, Jean and Shirley, along with Enid and a friend, 1953.

Sleeping in the Sun

My very best memory is of lying on warm hop bines listening to birds and bees and so many times being woken up to go back to the hut. Happy, happy days. What a pity they've gone.

Mrs M. Murray

Hottest Summer

I was evacuated when I was eleven so didn't go picking again until 1951 with my aunt and my six-month-old baby at Risebridge Farm, Goudhurst. 1976 was the hottest summer on record but when we started picking it rained non-stop for the whole three weeks. We were knee deep in mud. When picking was finished the farm was sold and the hops grubbed up. I still miss it. We went to the Kent Museum for Rural Life, Cobtree recently, and there were a lot of ex-hoppers telling a fund of stories. My mum's sister's ten children all went to Hazelden Farm, Cranbrook and still have their old hopping huts as holiday homes.

Irene Crimins

Delicious

I only wish I could take my grandchildren back, hop picking. What lovely memories! We lived in Wycliffe Road, Northfleet, when I was a child and hopping was our holidays. Early every morning we went to the corner outside the pub with all our neighbours and the mums sent all us children round the alley to the back of Brand's the bakers to get hot rolls as they came fresh out of the oven, then back to wait for our open backed lorry to take us out to Betsum for our day's hopping. We lit the fires for hot water for tea – it never tasted so good – and we'd have the rolls. When we children were not picking hops we went off to play and maybe pop into the orchard. We were told not to touch, but how could we not? What a shame it's all gone. Our children have missed so much: the friendship, the caring, the work. Everything about it was great and I'm glad I was there.

Mrs J. Watson

Worthwhile

It was one big holiday for us and the money we earned was spent on a good old-fashioned Christmas. Plenty of everything. The hop-picking was always a happy time and everybody helped each other.

V. Hammond

Hopping Evening

The other mums would ask my mum how she got me to sit and pick, as their kids just wanted to play. They didn't know I loved it. Before mum passed away we spent hours reminiscing. Still, when I'm sitting in the garden around early September, I'll say 'It's a real hopping evening' and I remember all the happy times.

Doreen Dillon

Finale

On the last evening we lit a big fire and sat around singing songs and remembering all the things that had happened. Everyone was great friends by then. We'd tell jokes and remember people who hadn't been able to come, then at the end we burnt all the old things we'd worn for hop-picking.

Evie Mortimer

Adventure

It's sad to think that things are not like that any more and that it's all over. Still, thank goodness we have memories. My only regret is that I can't do the same with my children and grandchildren. On the last day, all the families put their tables together, brought out all the food that was left over and held a party for the children. When we'd eaten we'd have a sing-song. I miss everything really: the getting ready to go, the getting there. Everything. To me it was a whole week

Irene Crimins' family ready for home.

Dolly, Joan, Len, John and Larry Thirkell at work.

George Turner recently revisited the hop huts and took this picture. His family slept and lived here throughout the picking season.

of adventure and fun, knowing you were going to see friends you had met each year.

Stan Dalton

Our Last Year

At the end of the week our parents queued up at the house office to draw their money or have a sub. Our last year we ever went was 1948 because the machines came in to do the pickers' jobs and it was really sad we weren't going to the picking any more because we were all happy people. It was like taking six weeks' holiday. We all came home looking brown because apart from the cold mornings we had lovely weather. One year around the seventies I found a calendar with a picture of Goudhurst. I had it enlarged and painted in 1973. It's still hanging in my lounge and always reminds me of where

David and I met in the hop gardens and the wonderful times we had together.

Doris Turner

All Changed

I regret my three boys didn't have the chance to experience the life, freedom and laughter we had. I loved the time there, but as mum was on her own and not in good health we had to pack it in. Fewer people we knew came as the families grew up and our last year we knew none of the people; all new. Also, they had day workers or home workers. They were locals who came in on lorries from the surrounding villages. There were eerie stories round the camp fires, sing-songs, the binmen's stories of the Notting Hill murders which happened during those years, and they used to tell us about them as they lived there. There

Mrs Fagan's two sisters. mother and father on a log by the fire.

was always the final big burn-ups of the straw from the mattresses before we left to go back home. The cold, the wet mornings, mud, the laughs in the fields – it was all part of it. I remember that last weekend: Mum and I came home on the train. It was so sad. Not like the year we came home by lorry and had a live cockerel in a string bag to fatten up for Christmas.

Enid Styles

Last Word

We went to the farmhouse to be paid off and there was a little old man asking for ten families to go and pick on his farm for him. So mum, gran and all my aunts decided as it was bad in London we'd do it and we all went off to Staplehurst. When we got there talk about 'One man went to mow'! He had one field and one oast house, one wagon, one horse, one row of huts, one nanny goat, one dog and one daughter. She did the measuring and tallying and her dad held the poke and drove the wagon. I'm seventy now, but that song Mary Hopkins sang 'Those were the days, my friend, I thought they'd never end' was true.

Rosina Amis

Glossary of Hop-picking Terms

Aisles or *Drifts*
Passages between each row of hops to enable workers' access to the plants.

Ale
An intoxicating beverage originally brewed from malt and not hops.

Alley Bodge or *Hop Dolly*
A purpose-built, sturdy, horse-drawn three-wheeled cart used for carrying manure along the aisles to feed hop plants early in the season.

Bagster
Before rams were used, bagsters trod down hops in a pocket to fill it to maximum capacity.

Bavin
A bundle of wood (S. Dalton).

Beer
An intoxicating, hop-based beverage.

Bine Dusting
Spraying hop plants with insecticides and fungicides while in the early growing stages.

Bine Hook
Curved metal knife on the end of an 8ft wooden shaft used by binemen for cutting down the tops of bines caught on the strings and wires.

Bines
Stems or vines of hop plants on which hops grow.

Binemen and Binmen
Workers who cut down bines using U-shaped hooks on long poles, delivering them to bins for pickers to strip off hops.

Bins or Cribs
Hessian bag bins strung along a collapsible wooden framework into which the hops were picked. They held about twenty bushels.

Brimstone Pan
Rimmed shovel used for placing cylindrical blocks of flowers of sulphur into the kiln fire.

Bushel
Capacity measure for dry goods such as hops, equivalent to eight gallons. A top picker could pick thirty bushels a day.

Bushel Baskets
Specially made wicker baskets used for measuring hops.

Cockle
Iron oven for heating the kiln.

Coir Strings
Tough, coconut fibre string used for supporting bines.

Cooling or Drying Floor
A long, upper room adjoining an oast in which hops were cooled after coming off the kiln floor. They were dried by constant turning with hop spuds to ensure even drying while allowing air circulation. This prevented hops turning mouldy and rotting.

Coping
Sewing across the top of full hop pockets.

Coping Twine
Extra strong twine for coping.

Cowl
The conical crown on an oast house roof was dual purpose: turned from the wind by its vane, it prevented backdraught while acting as a rotating smoke and hot air outlet from the hops roasting below on the kiln floor.

Crown:
A growing hop plant root.

Curing
Drying hops under controlled conditions.

'Dirty Picking'
Most pickers took pride in 'picking clean' but some slipped hop leaves and bits of bine into their baskets to add bulk. Tallymen discovering 'dirty pickers' marked their tally down. If repeated too often they were not invited to the farm the following year and had to find another farm.

Dryer
Worker responsible for the oast kiln and the drying of the hops. A specialist job.

Faggots
Variously: In town – small bundles of chopped wood sold for lighting fires. In the country – dry twigs and small branches used for firelighting and as the base of a hopper's bed on some farms.

Furnace
Originally wood-fired, later heated by charcoal which reached higher sustained temperatures. Modern kilns are oil-fired. While drying was in progress, kilnmen slept the night beside their furnace, knowing the vagaries of their particular kiln and alert to monitor its temperature, which was crucial to the drying process.

Green Hops
Hops freshly picked.

Green Staging
The slatted floor of the poke store allowed air circulation around pokes until ready for emptying onto the kiln floor. Good air flow helped prevent hops from contracting mildew, which rendered them useless.

Hair or Hairmat
Circular, horsehair mat spread across the kiln's upper floor slats to prevent hops from falling into the kiln, while allowing heat from below to circulate freely.

Heel
Hard-growing root of hop plant.

Hill or Stock
Hop plant, or group of shoots from a plant, built up into a mound above the normal soil level.

Hog
Slatted iron plate, angled to prevent sparks flying upwards from the fire and setting hops on the floor above alight.

Hop Dog
Strong metal claw with jagged teeth, usually with two wooden side handles, used for grubbing out hop poles.

Hop Gardens, Hopfields
Both names for the field in which hops are grown.

Hop Press
Vertical ram used to press hops down firmly into pockets. A hessian pocket was hung through a hole in the floor, supported by a strong leather strap, then fixed by its top edge inside the hole below the press. Hops were tipped into the pocket. The press was wound down to ram them tight, repeating until the pocket was packed solid, leaving only enough sacking at the top for the opening to be sewn up firmly.

Hop Spud
Extra wide, three-pronged fork, rounded and flattened to avoid damaging hops being turned on the cooling floor. The spud had an extra long iron shaft to its wooden handle to cope with the weight of damp hops.

Hop Varieties
Some of Kent's best known hop varieties are: Canterbury Whites, Farnham, Fuggles, Golding, Wye Target. Hops were also grown in Surrey, Sussex, Cambridgeshire, Hampshire, Herefordshire and Worcestershire et al.

Humulus lupulus
The hop plant, a hardy perennial. A member of the Cannabinaceae family of plants, a relative of nettle and hemp. Seedless, large hops are female. Smaller, pollen-producing hops are male and used for cross-fertilization. Only one or two male plants are planted to a field of female plants.

Kiln
Kiln designs varied, but basically had a meshed or slatted floor some thirty feet above a heat source kept at regulated temperatures controlled by the kiln man monitoring the temperature gauges and skilfully adjusting the fires accordingly. Fuel was wood, charcoal or oil.

Mechanical Hop-Picker
A Worcestershire invention. Machines were first used to pick hops in 1934 but not regularly used until the late 1950s. This marked the end of traditional hand-picking: the army of pickers 'invading' Kent every year were no longer needed. Machines only required three or four workers each.

Nidget
A horse-drawn, plough-shaped weeding and cultivating implement for use between the aisles.

Oasts, Oast Houses
Squat tower buildings with conical or square roofs surmounted with a movable, pointed wind vane. Walls were brick with a lath and plaster lining. The roof was clad with wedge-shaped clay tiles. Oasts housed the hop-roasting kiln. Square or round oasts depended on the architect's design.

Peeler or Pitcher
Long, bulbous, pointed iron bar used for making holes in the ground ready for hop poles.

Pipey Bine
Poorly developed tendril of hop bine. See Roguing.

Pockets
A 7ft long sack, stamped with name of the hop farm and date stamped according to the year to comply with Hop and Malt Marketing Board rules.

Pokes
Bushel baskets were tipped into pokes for delivery to the oast house. A poke was a large sack which held ten basketfuls of hops.

Poles or Bats
Poles, usually of chestnut or ash, approximately 25ft high, on which wires were strung to support hop bines.

Pole Pullers
Alfred May's pole pullers received 3s a day, starting work at 6 a.m. Each man pulled down poles with bines for eight bins, using a hop dog. Pole-pullers were charged 3d for every broken pole.

Pole Tug
A small horse-drawn cart used for transporting poles to hopfields.

Roguing
Pipey bines (under-developed hop stems) were pinched out (rogued) leaving more room and nourishment for stronger bines to grow.

Sample
To test the hops' quality the sampler used a keenly sharpened one- or two-bladed instrument for cutting out wedges of hops from randomly chosen pokes.

Scuppet
Shallow, wooden, three-sided box-like shovel with handle. Used to load dried hops into the hop press, known as scuppeting.

Set
A group of pickers working a drift or aisle, usually made up of a family or friends. Sometimes large families formed two or more sets, taking on a drift between them. Set also refers to a grouping of 100 hop hills.

Sprog
Forked stick.

Stains
Hop-picking left hands and arms badly stained. Hoppers' own remedies included scrubbing with coal tar soap or rubbing with soil or runner bean leaves.

Stiltmen
Stiltmen started their jobs months before the pickers arrived. Because of their height, poles were only accessible using stilts. Originally these were wood, but were heavy to manoeuvre in wet weather. They were later of hollow aluminium which was lighter for walking. Stiltmen tied coir strings to the top wires to train up the bines. In dry weather they erected lengths of netting to

protect bines from hot sun. Stiltwalkers needed ladders to reach and belt on the 13ft high stilts, fixing them round ankles and thighs with leather straps.

Stringing
A stringer hung strings from wires fixed across poletops to support bines. Poles had to be strung properly: loose strings offered inadequate support. There were several stringing methods, e.g.: Umbrella – strings radiated from a central pole to outside poles. Coley's Vinery system – vertical poles with an obliquely angled pole rising to the tops of the next row of bines. Butcher's system – strings went vertical to halfway up the pole, then were strung at an angle to top wires on the adjacent row. Each row shaded the one next to it. Parallel – strings were tied to the top wire and hung straight down to the ground. Worcester system – wires tied in to give a clear run through the aisles.

Sub, Subbing
Money given in advance off a worker's wages at the farmer's discretion. Some hoppers subbed all their earnings, going home with nothing for their efforts. Most saved carefully to buy children's clothes, boots or Christmas goodies.

Sulphur
Sticks of sulphur were added to kiln fires to impart a special flavour to beer and the yellow tint demanded by brewers. The practice ceased in 1980. Powdered sulphur was used as a fungicide in hopfields. Sulphur candles were burnt to kill off infestations of fleas and lice in huts.

Tally
Record of the number of bushels picked by individual pickers.

Tallyman, Measurer or Booker
Tallymen were important to both hoppers and farmers. To hoppers he meant the difference between good or average wages according to his fairness in the reckoning. Farmers expected good tallymen not to pay labourers over the odds. Alfred May's 'Regulations and Rules' stated that measurers had to keep order in hopgrounds, settling quarrels and 'disorderly conduct'; to see all regulations were followed; to 'attend properly and fairly to the pickers by filling his measuring basket full but as lightly as possible… without preference to any person and to take no half baskets except on Saturday.' Tallymen issued hop tokens, or recorded tallies by book, tally card or wooden tallystick, marking a notch on his and

the picker's tallystick for each bushel taken. Hoppers were paid according to the number of bushels recorded.

Tokens, Medals, Checks
Old forms of receipt for bushels picked. No money changed hands until the end of the picking season (unless it was 'subbed').

Tommy Hoe
A short-handled hoe for grubbing out weeds from around crowns.

Twilling, Training, Twirling, Twiddling
Twillers or twiddlers were usually women. New shoots on hop crowns were trained or 'twilled' up strings to give direct support. They required careful handling to avoid snapping off the easily broken growing tips.

Vane
Movable, pointed rudder with an extended tail, attached to a cowl. Vanes help regulate down draught to the kiln fires. Sulphur tinged the smoke yellow. Formerly wood, vanes are now mostly of cheaper fibreglass.